John Peter Schmitz

Human Physiology

Analysis and Digest, for the Use of Medical Students and Practitioners

John Peter Schmitz

Human Physiology
Analysis and Digest, for the Use of Medical Students and Practitioners

ISBN/EAN: 9783337140557

Printed in Europe, USA, Canada, Australia, Japan

Cover: Foto ©berggeist007 / pixelio.de

More available books at **www.hansebooks.com**

HUMAN PHYSIOLOGY;

ANALYSIS AND DIGEST.

FOR THE USE OF

Medical Students and Practitioners.

By Prof. John P. Schmitz, M. D.,

AUTHOR OF "OVER 3000 QUESTIONS ON LAWS OF THE HUMAN BODY;" AND
"THE CAUSE OF DIPHTHERIA AND THE DIFFERENCE BETWEEN
DIPHTHERIA AND CROUP."

SECOND EDITION.

PUBLISHED BY THE AUTHOR:

3321 Twenty-first Street, San Francisco, California.

1899.

PREFACE TO SECOND EDITION.

N. B.—A careful perusal of this Preface is especially
recommended to my readers.

The demand for a Second Edition of this text-book on physiology
within such a short time surely indicates that the book is much appreciated, especially when the Second Edition requires no additions,
alterations, or corrections of the main text of the First Edition.

It is gratifying to notice that the time for the study of medicine
in medical colleges has been extended to four years, which is surely
not too long for one who wishes to acquire something above medi-
ocrity in this science. In one respect, however, it may not be amiss
to allude here to a serious defect in some localities of not requiring
a better qualification in the knowledge of the Laws of Life.

Students intending to become physicians, have a right to demand
a most thorough teaching of the laws that govern the human body.
Deception in this regard is bad; and when diplomas are awarded to
those who have not a thorough knowledge of physiology, it endan-
gers human life and perpetuates conditions in the profession that
keep up the danger. It is the duty of all medical examining boards
to refuse licenses to applicants who are found deficient in this regard.

This work carefully distinguishes physiology from histology. It
is notorious that even in some first-class Colleges these two branches
are not unfrequently confounded. What can be expected from the
student when the teacher himself does not know that Physiology is
the science of the **Laws** of life and **Functions** of living organisms; Histology the science of minute anatomy. It is on account of
confounding these matters in text-books that the mischief is worked.
The student gets confused, and at last gets to hate and shun physiology, managing to cram a few dozen answers to questions in one or
two years in order to get out of it. The fact is, as the author has
always experienced, that, when physiology is truthfully and com-

prehensively laid before students, no branch of medical knowledge is so pleasing and fascinating. The student soon gets to feel internally that the true and thorough knowledge of the **Laws of Life** is the very **Foundation** of medical competency, diagnosis, and safety of treatment.

The conscientious student will ask himself: How shall I be able to make a sure, true diagnosis? How shall I be able to know what remedy to administer? How shall I be able to know the physiological actions of remedies? The books on *Materia Medica* and *Practice* tell me that the action is such and such, but not one tells me *how the remedies act*. How shall I be able to know whether what I observe in a patient is to be attributed to remedies or to the disease?

A common sense man says, give me a lawyer thoroughly versed in the law and I will trust my case to him. Can a lawyer ignorant of the law know when he does harm to his case? Certainly not. It is exactly the same with the physician. The practitioner, ignorant of the laws of human life, may be not inaptly compared to a blind bird; it **may** find the grain on the way-side, but the chances are against it.

The greatest responsibility falls on the medical college faculty or officers, who supply the chair of physiology with simply a bookworm. In fact, it seems very often as if any physician is thought competent to fill that chair as Professor, provided he is able to talk the hour away two or three times a week. Such a Professor cares little whether the student has fully understood him or not, or in fact whether he has understood him at all: for, if he did he would not allow students, as very often happens, to cram a few dozen questions and answers for what is termed "Examination," and then dispense with physiology for the balance of the College course.

It has often been publicly remarked, that no two physicians agree in regard to treatment. Why is this? Because if all physicians were thoroughly educated in physiology they would know positively what is required, and there would be no difference of opinion in treatment. Then the treatment of human beings would be truly scientific, but it cannot be without that knowledge. Physiology is no longer guess work; it is as positive as any other science, and the physician who does not know the Laws of life is either at fault himself or the blame falls back on his teacher. It is not due to the science.

The author for years has maintained that the time required for the study of physiology in the medical colleges is too short, and he has kept his students for the full term of four years of college study on physiology. That he was not alone in this idea it is agreeable to notice that the College Faculty of Physicians and Surgeons, **London,** have by resolution extended the time for the study of physiology of three years, so that hereafter students are required to continue this subject for the full term of five years.

Anyone who maintains that medical students can be excused from the study of physiology before the end of their college life, knows but little about it. If he comprehended this science, he would know the importance of that study.

Does the Professor on physiology ever reflect on the following:— Am I fit for the position? Do I do my duty? Am I not neglectful? Do I see that every student under my care fully comprehends the subject? Will I not be partly responsible for the acts of a physician once in my care who does more harm than good to patients, or perhaps shortens their lives on account of not having received from me thorough teaching?

On no other chair in medical colleges does the responsibility so forcibly rest as on the chair of physiology; because a doctor once out of college can easily continue the study of other branches of medical knowledge, but not that of physiology. This latter has to be acquired in the college, because to fully comprehend the laws of life, requires a thorough and practical professor to teach, to read, to explain, to illustrate a subject sometimes in many different ways, before every student fully comprehends it. Besides, the explanations and illustrations have to branch off more or less on a subject of another chair, or to several at a time, so as to make the point understood.

The so-called **Quiz-books** in the market, *i. e.,* books containing questions and **Answers** on physiology, are surely detrimental to the student in acquiring a knowledge of physiology. No man can learn any science out of quiz-books.

Concurrently with this text-book, the author has prepared a small book (paper bound) entitled *"Over 3000 Questions on Laws of the Human Body."* (See page 348.) That book contains **Every Point** in this text-book in question form—this work the **Answers** in detail, and which follow one another similar as the questions. The student, therefore, fortunate enough to possess these two books, can, with more ease and facility understand and truly comprehend

the laws governing all the organs of the human body. It is, if I
may so express it, a chewing up of the scientific food for more easy
digestion by the student.

It may not be too much to assume that the author of this work is
the first who ever laid down in a medical college text-book, the true
fact of the following physiological laws:

First—That the **Stimulus** for respiration and circulation are the
carbonic acid elements of the venous blood to the heart and lungs.

Second—That the **Acid** for the gastric juice normally originates
in the lower portion of the œsophagus.

Third—That **External Sensation** lies in the sense-organ.

Fourth—That the living body comprehends an **Immaterial Vital Principle** or **Soul**.

Fifth—That all brain and spinal-cord **Nervous-centre Actions,** voluntary or involuntary on external organs, are due to
Reflex Actions only.

Sixth—That the **Cause of the Capillary Circulation** in the
animal body is peristaltic.

Seventh—That the defective mucous membrane is the **Primary
Cause of Consumption.**

Eighth—That the physiological action of remedies will become an
exact science as soon as physiology is truly comprehended, and not
before.

J. P. Schmitz, M. D.

3321 Twenty-first street,
   San Francisco, Cal.

PREFACE TO FIRST EDITION.

A new book on Physiology——with such works as those of Landois, Dalton, Yeo, Kirkes, Chapman, Forster, Flint, Huxley and others before the public, would appear almost superfluous; yet this work may demonstrate that there are many points which even now require further elucidation.

In any exposition in regard to physiology, it is difficult to arrange the subject-matters both in the order of their relative importance and of their logical sequence. In the following pages, however, we have endeavored to set forth in a brief yet clear and sufficiently extensive manner the more important points of physiological interest. The text is illustrated with many original diagrams. In preparing these illustrations no effort has been made to *picture* organs and their functions; they are merely designed to illustrate the text.

Where it has been deemed advisable, sufficient anatomy is given to place the subjects in a clear light before the reader, without the necessity of reference to special text-books. This plan may assist even the physician in recalling forgotten parts of structures and of organic functions. Furthermore, it is believed that this work, in its explanations of stimuli and sensation, will furnish the key to a better understanding of the physiological actions of remedies.

The reader's attention is especially called to the *primary causes* of respiration and circulation, the origin of the acid of gastric juice, sensations, the cause of capillary circulation, functions of nervous centres, mucous membranes, and albuminoids, as set forth in this work.

The science of physiology presents a well-nigh inexhaustible field of investigation, requiring careful study and deep thought. No one physiological subject can be singled out and an intelligent judgment passed upon it separately; the laws of life must be understood as an entirety.

A glossary containing over 1000 technical medical terms, properly defined, has been appended, so that any educated person can understandingly read the book without the assistance of Latin and medical dictionaries.

An exhaustive index of all the subject-matters in the book is also appended, so that any subject may readily be found in the text without loss of time. This may be an agreeable feature to the physician in refreshing his memory on special organic functions. The scientific physician cannot dispense with the book on the laws of life from his library table, any more than the lawyer can dispense with the book on the laws of the state from his table. The belief that after having once passed successfully the medical college the study of physiology is no further necessary, is surely a great mistake. It is hoped this work will prove not only helpful to the student, but also stimulate him to acquire a *thorough* knowledge of the organic physical and physiological functions, and at the same time render his study a pleasurable one, on account of the condensed yet clear and comprehensive manner in which the subjects are treated. The scientific treatment of abnormalities of organic functions of the body can be accomplished only through a thorough knowledge of the laws of life.

The many medical students who, during their three-years course in the medical college, have listened to my lectures and recitations on physiology, are very gratefully remembered for their marked attention and gentlemanly bearing during the lecture hours.

San Francisco, California,
     May, 1894.   J. P. Schmitz.

EXPLANATORY.

Stars indicate that the subject-matters are illustrated; the illustration may or may not be on the same page as given in the *Contents*. A separate and complete list of the illustrations, alphabetically arranged, will be found on page 19, *et seq.*

All technical words in this book will be found, properly defined, in the *Glossary* at page 277, *et seq.* Also, a complete *Index* will be found at page 305, *et seq.*

CONTENTS.

INTRODUCTION.

CHAPTER III.

CHAPTER IV.

CHAPTER V.

CHAPTER VI.

CHAPTER VII.

CHAPTER VIII.

CHAPTER IX.

CHAPTER X.

CHAPTER XI.

CHAPTER XII.

CHAPTER XIII.

CHAPTER XIV.

CHAPTER XV.

CHAPTER XVI.

CHAPTER XVII.

CHAPTER XVIII.

CHAPTER XIX.

CHAPTER XX.

CHAPTER XXI.

CHAPTER XXII.

CHAPTER XXIII.

CHAPTER XXIV.

CHAPTER XXV.

CHAPTER XXVI.

CHAPTER XXVII.

ILLUSTRATIONS.

PHYSIOLOGY; ANALYSIS AND DIGEST

INTRODUCTION.

Physiology is the science of the laws of life and functions of living organisms.

An **Organism** consists of a combination of organs, and has specific functions. In structure it is capable of performing actions and producing effects not only by itself and within itself, but also on matter external to and outside of itself. An **Organ** is a part of an organism, and its action is its function.

The organic world includes both *animal* and *vegetable life*, with their component and physical properties, chemical composition, and vital phenomena. Animal as well as vegetable life depends upon the action of its individual organs, distinct but mutually combined and dependent.

Organic bodies differ from inorganic by the introduction, assimilation, combination, and reconstruction of new or fresh matter. Certain inorganic substances aggregate to themselves fresh material, enlarging in size and quantity, thus showing a quasi-assimilation; but they do so only by the addition of particles of matter to their exterior. The organic living structure grows by the addition of new matter not only to its surface, but throughout its entire mass; and at the same time it continually changes—decay and repair going hand in hand. As soon as the mutual dependence of the different organs in an organized body ceases, it becomes like an inorganic body, the only force that keeps it in form being cohesion.

The vegetable or plant, through its *chlorophyll*, or green coloring matter, decomposes the carbonic acid, ammonia, and water, which were absorbed by its roots and leaves, and utilizes them as food. There is a chemical change effected under the influence of solar light, the carbon of the carbonic acid becoming fixed in the structure of the plant, and the oxygen exhaled. Kirkes states that "animals cannot thus use inorganic matter, and never exhale oxygen as a product of decomposition."

Organisms differ widely, except in cellular assimilation for nutrition and repair. In their tissues the vital phenomena are performed by living cells, acting independently in response to stimuli; and for proper action require a certain degree of warmth and moisture, without which chemical interchanges cannot proceed, the organism becoming either inactive or destroyed.

As this work is designed to elucidate the principles of **Human Physiology** only, the subject-matters of the following pages are, therefore, restricted to this branch of physiological science, as far as it is deemed compatible with fullness of elucidation.

About fifteen of the elements known to chemists take part in making up the tissues of the human body, the majority being present in small and varying quantities only. Four elements, however—hydrogen, oxygen, nitrogen, and carbon—occur always in large quantities, constituting 97 per cent of the animal frame—hydrogen, oxygen, and carbon being the most constant and abundant.

The human body, anatomically, is divided into four parts—head, neck, trunk, and extremities, each of which contains many different organs; viz., in the head: brain, eyes, ears, etc.; in the trunk: heart, lungs, liver, kidneys, spinal cord, etc.; in the extremities, and body, every bone and muscle constitutes a distinct organ.

A **System** consists of a combination of several organs, similar in texture, but scattered throughout the body—as the arterial, muscular, osseous, and nervous systems.

An **Apparatus** is formed by the combination of entirely different organs; for instance, the circulatory apparatus is formed by the heart, arteries, capillaries, and veins.

By **Tissues** we understand the texture or grouping of two or more different anatomical elements interwoven into each other, of which any part of the body is composed.

An **Anatomical Element** is the smallest natural division of the organism. If the anatomical element—*e. g.*, a blood corpuscle, muscular fibre, or an epithelial cell—be separated from the organism, and further broken up, it loses its identity and function. The union of these elements constitutes the tissues; consequently, all parts of the human frame are made up of anatomical elements, forming organs, systems, and apparatuses connected with each other. Each anatomical element, wherever situated in the body, possesses the property of becoming excited by a stimulus proper and essential to it. Herein lies the great secret of vitality or life.

A **Cell** is a nucleated mass of protoplasm, in the form of a small vesicle, composed of a jelly-like or somewhat fatty substance. In all kinds of tissues the function of cells differs; some stimulating, while others secrete, absorb, repair, or reproduce. A cell is the first noticeable formation to propagate vital phenomena. Its function is not only assimilative, enabling it to increase in size, but also contractile, originating vital energy within itself and transmitting it to other tissues. That cells differentiate by a distinct investing envelope, termed cell-wall, or cell-membrane, has not been confirmed. **Cell-reproduction** is by *fission or division.* Each cell is an entity. In reproduction, its tissue, nucleus, and nucleolus (*protoplasm*) elongate and divide into two halves; each forming a complete cell.

CHAPTER I.

NUTRITION.

Nutrition is the assimilation of nourishment in the body. The nutritive substances are derived from proximate principles already existing in animal and vegetable foods, which are adapted to enter into the composition of the different parts of the body. A *proximate principle is a compound of elements existing in the animal or vegetable, solid or fluid, and which can be extracted* without altering or destroying its properties. The most necessary of these for nutrition are:—

1. *Albuminous (Nitrogenous)*, of organic origin, and derived from both animal and vegetable foods.

2. *Starchy, Saccharine, and Oleaginous (Non-nitrogenous)*, also of organic origin, and derived from both animal and vegetable foods.

3. *Inorganic (water* and *saline matter)*, derived from liquids and solids.

The Coloring Matter, consisting of nitrogenous compounds, is believed to originate in the body.

Albuminous, or **Nitrogenous,** substances are of great importance for animal tissues and fluids, no food being nutritive without them. Albuminous matters are called nitrogenous proximate principles because they contain nitrogen in addition to carbon, hydrogen, oxygen, and sulphur. Owing to the nitrogen contained, they differ from the non-nitrogenous (starch, sugar, and fat). The principal food substance of muscular tissue is its albuminous matter, and musculin, similar in composition. *Albumen* exists naturally in the most pure state in the "white" of an egg. *Albuminous* matter refers to animal and vegetable food containing some albumen mingled with other nutritive substances; while *Albuminose* is the converted albuminous matter, liquefied in the stomach; it is capable of passing through animal membranes, which pure albumen cannot do. Albuminous substances are hygroscopic—*i. e.*, they contain moisture—and may, therefore, on evaporation of the moisture, become solidified; however, if brought again in contact with moisture they will absorb liquid and swell, but never go beyond the weight or quantity originally associated with them. In addition to being hygroscopic, the albuminous proximate principles coagulate, but when coagulated cannot resume their original condition. In regard to coagulation, the peculiar chemical characteristics of albumen, fibrin, and casein are distinguishable as follows: Albumen coagulates when boiled; casein, when in contact with an acid; fibrin, when gradually withdrawn from the blood-vessels and exposed to the air. When albumen coagulates, its properties are permanently changed; it assumes a solid form, but retains the same quantity of water as before. Vegetable albumen exists in all vegetables, and forms their most essential constituent when used as food. In one thousand parts of human blood there are about 75 parts of albumen, which, by its presence in the blood, has the physiological property of mutually exciting in other proximate principles catalysis, or catalytic transformation—*i. e.*, it dissolves, destroys, or changes them into other forms. The albumen, too, uses itself up during these transforming processes.

Non-nitrogenous proximate principles are divided into two groups—carbohydrates (starch and sugar), and hydrocarbons (fats); the latter including all varieties of oleaginous matter.

Starch.—During mastication starchy matter is partly converted by the saliva into glucose, which has during digestion the same significance as sugar. Starch, sugar, and fat are termed non-nitrogenous because they contain no nitrogen, but consist of carbon, hydrogen, and oxygen—carbon forming 44 to 84 per cent. Starchy substances are converted into sugar by the pancreatic juice also, during digestion. The necessity and desire for vegetable food is so great that restriction to a diet of animal food would surely in time prove fatal to the human body. Saccharine and oleaginous substances are necessary, and found together in most vegetable food; but neither is alone sufficient for nutrition. In a short time health would fail without the albuminous and fatty proximate principles.

Sugar.—The sugar in the food, as well as that formed in the liver, disappears by decomposition in the animal fluids; normally, it does not appear in any of the excretions. Sugar contains much carbon, which may be demonstrated by placing a small quantity of sugar on a hot stove; the water evaporates and the carbon remains; but the sweetness of the sugar is destroyed, as its molecules are split up, the atoms of hydrogen and oxygen having separated from the carbon. *Sugar of milk* and *liver sugar* originate in the body; the cells of the mammary gland and those of the liver, respectively, having the function of effecting the union of the proper number of atoms of carbon, hydrogen, and oxygen into molecules of sugar. Liver sugar is found in the substance of the liver and in the blood of the hepatic veins.

Fats.—As a general rule, the oleaginous matter in animal substances is found in distinct masses or globules suspended in the serous fluids; interspersed in the interstices between the anatomical elements; in the interior of cells; or deposited in the substance of fibrous membranes. (Even in vegetable tissue, oil is always deposited in distinct drops or globules.) Fat or oily matters are necessary ingredients of food; they cannot be adequately substituted by starch and sugar. The production of fat in the body requires a supply of proper materials. Sometimes in the animal economy, when emaciation takes place, the oil partly disappears from the cavity of the adipose vesicle (fat-cell) and a watery serum fills that portion of the cell, but the serous and oily fluids remain distinct in the different parts of the vesicle. In milk the oily matter is in larger globules and in greater quantity than in chyle. In cow's milk the oil, or milk-globules (Fig. 1), have a pasty consistency, on account of the larger quantity of palmatin in proportion to the olein, and

Fig. 1.

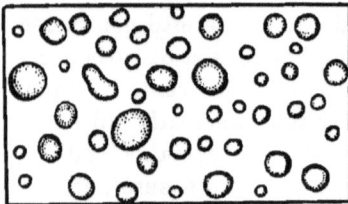

Oil Globules in Cow's Milk.

when churned their combination forms butter. The fat of the body is partly taken in with food and partly formed in the body by the metamorphosis of other (not fatty) proximate principles; possibly, also, by the transformation of starch and sugar, as starch disappears from oil-seeds when ripe. The important physiological fact in relation to amylaceous, saccharine, and oily substances, as to their source and final destination, is that they are of organic origin, appearing first in vegetable growths. They are also derived, to a certain extent, from other organic materials in the bodies of animals, and continue to be formed when no similar food has been taken. The fixed oils, of either animal or vegetable origin, consist of three oleaginous ingredients— olein, stearin, and palmatin. The *olein* is the liquid portion and the distinctive proximate principle, and is found mostly in animal oils or fats; the other two, *stearin* and *palmatin*, are more solid ingredients. Stearin is most abundant in tallow and other animal fats; palmatin, principally in palm oil, cocoanut oil, human fat, and butter. Fat when taken with food is not altered in the stomach, except that it is freed from its enclosing membranes, and, entering the duodenum, the greater part is then converted into a fine emulsion under the influence of the pancreatic juice. It is then ready to be absorbed by the lacteals, and enter the circulation, where its essential characteristics are destroyed through a series of changes. The transformation of fat in the vital economy evokes great activity of atoms and molecules, thereby producing heat. Fat has a low power of conducting heat, so that the fat deposited under the skin serves to retain the heat produced in the body. The further use of fat in the vital economy is to fill up spaces, and, as a soft, yielding material, to serve either as a bed for delicate structures or as an envelope for them. In health a moderate quantity of fat is appropriated by the organism, but, when introduced with the food in greater quantity than the absorbents are capable of taking up, then the surplus passes out with the feces and a small quantity with the sebaceous secretions.

Starch, sugar, and fat, after serving the purposes intended in the body, become transformed and decomposed, and are finally expelled in the form of carbonic acid and water.

Inorganic Proximate Principles are compounds of chemical elements. Those most commonly found in the body are water, sodium and potassium chlorides, phosphates, sulphates, carbonates, lime phosphates and carbonates, and magnesium phosphates and carbonates. The alkaline carbonates, phosphates, and sulphates are formed partly within the system during the decomposition of organic matter.

Water is not changed in the body, but absorbed, circulated, deposited, liberated, and expelled unchanged; though it remains necessarily present in sufficient quantity. It enters into the composition of every fluid and solid tissue of the body, and the system suffers more quickly from want of liquid than from want of solid food. Water is a very important and essential ingredient, because it holds the particles of solid material in fluids in solution, facilitates their endosmosis and exosmosis, enables nutritious elements to find their way into the blood, to penetrate the substance of the solid organs, and permeates every part of the body. It brings the organic and inorganic materials in contact, thereby enabling them to assume new forms. Through its agency chemical changes are accomplished in the various organs; for even the most solid parts of the body contain water, which, when evaporated, reduces their weight. About sixty per cent of the entire weight of the body is water, derived from without either in drink or food. About 2000 grammes (nearly four and a half pounds) of water per day is introduced into the system of an adult. Water passes off with. the urine, feces, and by the lungs and skin. In 100 parts of water passing from the body, the lungs give off by exhalation about 20 parts; the skin, by perspiration, 30; the kidneys, with the urine, 46; the other 4 parts passing off with the feces, thereby rendering an almost continuous supply of water necessary.

Lime Phosphate is contained in all animal and vegetable foods, and all animal solids and fluids of the body hold it as a constituent element. In quantity it ranks next to water, and is present in greater quantity than any other mineral salt, forming more than one-half of the substance of bone. It is not soluble in water, but is held in solution in the blood by albuminous substances. The lime phosphate imparts rigidity to the solid tissues, the stiffness or hardness of bones and teeth being due to its presence. It may be extracted by macerating bones in dilute muriatic acid, after which, while the organic material retains its form, the bones are, however, capable of being flexed or twisted. The peculiarity of lime phosphate is its combination internally with animal matter of all kinds of tissues, similar to the manner in which the coloring matter unites with the other ingredients in colored glass, the lime phosphate still retaining its original character and composition. If in childhood ossification does not properly proceed on account of the insufficiency of lime phosphate, rhachitis, or rickets, results, and the spine and bones become distorted. In adult life an insufficient supply of lime phosphate causes morbid softening of bones, termed *osteomalacia*. In urine, lime phosphate is held in solution by acid bi-phosphate. Whenever urine is rendered alkaline by the addition of soda or potassa, the earthy phosphates, of which lime phosphate is one, are precipitated in the form of a turbid white sediment.

Lime Carbonate, like lime phosphate, is found in all animal solids and fluids, its proportion compared with lime phosphate being about one-seventh by weight. The carbonate is held in solution in fluids by virtue of the presence of alkaline chlorides, such as the sodium and the potassium chlorides.

Magnesium Phosphate, like lime phosphate, is also obtained from food, and found in all the tissues and fluids of the body; a large quantity existing in the muscles, and much more, comparatively, in the brain.

The **Alkaline Salts,** such as carbonate of soda and potassa, are essential to nutrition; yet, unlike mineral salts, they are· not usually introduced in their own forms but rather with fruits and vegetables containing salts of soda and potassa, combined with various organic acids, such as citrates, malates, and tartrates, which are decomposed and become alkaline to animal secreted juices and fluids of the body, and may be increased by an exclusively vegetable, or lessened by an exclusively animal, diet. Muscular fibre and animal tissue contain phosphates, while vegetable matter abounds in said organic acids, the latter, by decomposition in the system, producing carbonates of the same base. If the food be both animal and vegetable, the salts are then found to be in proper proportion. As already stated, alkaline salts are derived from food, and excreted with perspiration, mucus, and urine. An animal diet increases the quantity of tissue-acid (a form of lactic acid) in the system, and consequently more acid (though changed) is then found in the excretions.

Sodium Chloride (table salt) is, next to lime phosphate, the most abundant of all mineral constituents in animal tissues and fluids. It has the property of reg-

Fig. 2.

ulating the phenomena of endosmosis and exosmosis, or the transudation of nutritive fluids through organic membranes. Salt excites the digestive fluids. Moderately used, it is absorbed in the alimentary canal, passes into the blood, then into the tissues, returns to

Crystallized Sodium Chloride.

the blood, and is finally discharged with the urine, mucus, and cutaneous perspiration. Of the sodium chloride in the adult system more than ten per cent passes off daily, amounting to about fifteen grammes, or a little less than four drachms, of which about thirteen grammes pass off with the urine and two with mucus and perspiration. Sodium chloride is very important in nutrition. This, even the wise farmer knows, as is shown by his liberal supply of common salt to cattle to improve their condition. It furnishes the principal chemical material for the production of the hydrochloric acid of the gastric juice.

Milk.—Nature provides the only perfect combination of food elements, in the form of mother's milk, to sustain and nourish the infant. Milk contains all the important groups of nutritive substances, consisting of albuminous matter, casein, sugar, and inorganic salts, especially a large quantity of lime phosphate. The fat globules suspended in milk (Fig. 1) give it an opaque white color. As the casein and albumen are associated together, milk does not solidify by boiling, but is covered by coagulated albumen, the casein remaining liquid. Any dilute acid precipitates the casein and curdles the milk. A moderately warm atmosphere transforms the sugar in milk into lactic acid, or atmospheric electricity may cause a similar change.

Bread.—Bread is the most nutritious of all combinations of vegetable matter. It is deficient in fat; hence, butter, fat, or some other form of oleaginous food is usually eaten with it. The flour of cereal grains, of which bread is made, contains starch, albuminous matter, and mineral salts. **Wheat** is the best grain for bread, on account of its large proportion of albuminous (nitrogenous) matter. It contains also a peculiar adhesive or fibrinous substance called *gluten*. **Oats,** next to wheat, contains the largest proportion of albuminous matter. It contains also an indigestible vegetable substance, or cellulose; consequently, thorough boiling and straining, to remove the cellulose, make it more digestible. **Indian Corn** is especially rich in fatty ingredients. **Rice** contains much starch, but very little albumen or fat.

Meat.—The muscular flesh of various animals is extremely nutritious as food for the human body, containing in every hundred parts about 78 of water, 15 of albuminous matter, 5 of fat, and 2 of mineral salts. The minerals consist of sodium and potassium chloride and phosphate, lime and magnesium phosphate. Meat loses, by any method of cooking, from 25 to 30 per cent of its weight, owing to the escape of water and liquefied fat.

Eggs.—The yelk of the egg contains in every 100 parts about 52 of water, 30 of fat, 16 of albuminous matter, and 2 of mineral salts; the white of the egg contains in every 100 parts about 78 of water, 20 of albumen, and 2 of mineral salts. The mineral matter consists mainly of sodium and potassium chlorides, potassium sulphate, lime and magnesium phosphates.

Of vegetable foods, some are valuable to the animal system on account of their starchy and albuminous ingredients, others for their saccharine and watery juices, while the green parts are doubtless useful because of the iron contained in their chlorophyll.

The value to the system of any food substance depends upon the amount and *requisite combination of the different proximate principles contained* — but not upon any one singly, even if taken in large quantities. The quantity and kind of food required daily varies with circumstances— such as the size and weight of the body; development of the muscular and other systems; temperature; and especially the amount of physical activity. More food is required in cold than in warm weather; more by a person of muscular than of adipose or phlegmatic constitution, and more in activity than in repose. Carbon and nitrogen are considered important constituents of efficiently nutritious food; carbon forming the most abundant and characteristic ingredient of all organic combinations, while nitrogen is the distinguishing element of albuminous substances, constituting one of the elements of all albuminoids, such as the alimentary digestive secretions. Of these two (carbon and nitrogen), about seven ounces (or about 224 grammes) of carbon and five drachms (or about 20 grammes) of nitrogen are continually required to support the system of an adult.

During sleep or fasting, some of the stored-up molecular energy constituting already formed tissues is used in the economy; therefore, this energy has to be renewed by introducing nourishment into the system; otherwise, the vitality required to keep up organic functions would soon diminish, and the organism begin to emaciate. On the other hand, if more energy than the system absolutely requires is supplied, then, from its superabundance the organism is in some way or other thrown out of order. This happens when food is introduced with which either certain organs or the entire system is already saturated, for then the functions of the organs become deranged.

The **Amyloid Compounds,** such as glucose, glycogen, and dextrose, differ from protein compounds by containing neither nitrogen nor sulphur.

3

CHAPTER II.

ACTIVE PRINCIPLES IN LIQUIDS AND SOLIDS.

The most active principles in the liquids and solids of the body are the albuminoids, viz.:

Fluids — Casein, ptyalin, pepsin, pancreatin, mucosin, and myosin.

Solids—Collagen, chondrin, elasticin, and keratin.

In **Coloring Matter**—Hæmoglobin, melanin, bilirubin, biliverdin, urochrom, and chlorophyll.

Casein is the principal nutritive substance in milk. When solidified it forms cheese. Boiling does not coagulate it, but dilute acids and magnesium sulphate do. Casein, when subjected to a highly acid condition of the gastric juice, or at a high temperature, coagulates. Milk, therefore, is not a proper article of diet when the stomach is abnormally acid.

Ptyalin is the principal active agent in saliva. It augments the stimulation for increased secretion of saliva, which converts hydrated or moist boiled starch into glucose. The daily secretion of saliva in an adult is about 40 fluid ounces.

Pepsin is the principal constituent of the gastric juice. It is secreted by the glandular follicles of the stomach to aid digestion. About fifteen parts of pepsin are found in one thousand parts of the juice. During digestion the pepsin converts the albuminous (nitrogenous) matter into peptones. The daily secretion of gastric juice by an adult is about six pints. (*Vide* Digestion.)

Pancreatin is the characteristic active ingredient of the pancreatic juice. Its proportion is about ninety parts of pancreatin in every thousand parts of juice. The amount of the daily secretion of juice by an adult is about ten ounces. The physiological property of the alkaline pancreatic juice is to emulsify and partly saponify fatty matter and to neutralize the fatty or oily acids. The juice has also the power of converting starch into glucose, even more so than saliva.

Mucosin, or **Mucin**, is a viscid, thick, glutinous constituent of mucus. It gives to mucus its glutinous consistency. There are

a variety of secretions called mucus that have the peculiar physical character of viscidity, keeping the membranes moist, thus enabling them the better to perform their functions.

Myosin is the contractile substance of muscular fibres.   After
death it coagulates; then the muscle loses its contractility and assumes a cadaverous rigidity.   Cold retards its coagulation.   It is
completely soluble in dilute acids and alkalies; on this account
abnormally increased acidity or alkalinity of the system produces
muscular weakness.

Collagen forms the homogeneous interstitial mass of bones,
periosteum, tendons, ligaments, fasciæ, and connective tissue.
When boiled it produces gelatine or glue.   One part of collagen is
precipitated in 5000 parts of water if tannic acid be mixed with it.
Tannic acid hardens the collagen of the fibrous tissues in hides.
They then become less permeable to water and almost incapable
of putrefaction—leather.

Chondrin is the organic basis of cartilage.   It is jelly-like in
consistency.

Elasticin is the homogeneous material of tissues.   It is found,
for example, in the middle coat of the larger arteries, elastic ligaments of the spinal column, ligamentum nuchæ, and all other yellow elastic structures.

Keratin is the resisting and indestructible substance of the
hair, nails, epidermic scales, feathers, and all horny tissues.   Sulphur is one of its constituents.

The **Coloring Matters** are all albuminoid compounds.   The
red coloring matter (hæmoglobin) is the most abundant, being the
agent by which oxygen is absorbed in the lungs, taken up by the
blood, and distributed in the organism.   Animal coloring matters
are found in the blood, in the blackish-brown solid tissues, bile,
and urine.

Hæmoglobin is the coloring matter of the red blood-corpuscles, constituting about 90 per cent of their bulk.   It is a powerful
absorbent of oxygen, retaining it in loose combination; it contains
about 0.42 per cent of iron, derived from vegetable and animal
foods.

Melanin is the dark coloring matter of tissues, found in the
choroid coat of the eyeball, in the *rete Malpighii* of the skin, in the
hair, and is more abundant in black and brown colored people,
producing different grades of color according to its proportion in
the tissue.

Bilirubin is the reddish-yellow or orange coloring matter of the bile. It is formed in the substance of the liver from the hæmo-globin of the red blood-corpuscles dissolved in the spleen. (*Vide* Spleen.)

Biliverdin is formed partly by transformation from bilirubin and partly from chlorophyll. It is the green coloring matter of bile. When animals feed on green vegetables the coloring matter of their bile is extremely green.

Urochrom is the coloring matter of urine. Fresh, normal urine has a light-yellow or amber color.

Chlorophyll is an alkaloid constituting the green coloring matter of plants. It is produced in the vegetable cells. If defi-cient in plants they grow up blanched, or in a chlorotic condition. Its essential substance seems to be iron. The entire process of veg-etation, production and accumulation of organic material in the form of starch, sugar, cellulose, woody fibre, and the substance of vegetable tissues, is inseparably dependent on the presence of chlorophyll. Its function is not that of a simple chemical reagent, but of an active constituent, its presence being just as essential as the presence of air, sunlight, warmth, and moisture to the living vegetable organism.

Urea is a colorless substance, with a neutral reaction, soluble in water, and crystallizing in four-sided prisms. It forms the prin-cipal solid constituent of urine. Formerly, it was regarded as an organic base or albuminoid of animal origin; though of late it is placed amongst the amides, and termed carbonic oxide, carbonyl, or amide of carbonic acid. It is found not only in urine, but in such excreted fluids as perspiration and tears. That urea, as such, exists normally in the blood and tissues, must remain a matter of doubt. It is an established fact that some persons, when eating meat in sufficient quantities, have excreted as much as 1540 grains of urea in twenty-four hours, yet apparently enjoyed perfect health. Now, if, within a period of twenty-four hours, this amount of urea existed already formed in the blood and tissues, it would poison the person and produce unconsciousness. Hence, we are con-strained to conclude that urea is completely formed only by the organs of exit. The blood, lymph, and tissues contain urea ele-ments. They are secreted and urea completely formed by the cells of the uriniferous tubules of the kidneys, and by the sudoriparous cells of the skin, and are by these organs (kidneys and skin) ex-creted. When some of the excretory channels are in a defective condition, the ready formed urea is easily reabsorbed by the capillary veins and capillary lymphatics, and it may then be carried to any part of the body, and be found in the blood, lymph, and tissues. Urea constitutes about 26 parts in each one thousand parts of urine. About sixteen parts of urea may be found in each hundred thousand parts of blood. About 33 grammes (nearly 500 grains) per day are excreted by an adult. A larger quantity is produced during waking than sleeping hours (accounted for probably by the action of both the nutritive and digestive functions), and more is produced in the later than in the earlier part of the day—mostly discharged from 6 to 10 P. M. The quantity excreted varies in proportion to the amount of albuminous and starchy matter in the food. As urea is now placed amongst the amides, starch must, consequently, contribute some elements to its formation. Urea is increased by a diet of animal food, also by unusual muscular exertion, diminished by vegetable food, and reduced to its minimum by a diet of non-nitrogenous matter, such as sugar and fat. It is not dependent solely on food, however, but is partly derived from the metamorphosis of the tissues of the body, as some urea is discharged even when no food is taken. (*Vide* Urine.)

Uric Acid is next in importance to urea, and is found in urine. It is a colorless, crystallizable substance, soluble in either cold or hot water, and, like urea, formed in the organs of exit, either from substances not properly formed into urea, or from elements arising from the metamorphosis of albuminous and amyloid substances of the food and of the ready formed tissues of the body. Its principal bases are acid urates of sodium and potassium, the sodium urate predominating. Animals nourished principally by vegetable food excrete more uric acid than urea, while those subsisting mainly on albuminous (nitrogenous) diet excrete less uric acid but more urea. The quantity excreted by a human adult is about from six to nine grains in twenty-four hours, or in the proportion of one of uric acid to forty-five of urea. Urinary gravel and calculi are often composed of uric acid.

Urine, on account of the great quantity of salts contained therein, is not absorbed from the bladder, as strong salt liquids are not osmotic. Bathing in fresh water relieves thirst, but bathing in strong salt water does not.

CHAPTER III.

THE MUCOUS MEMBRANE AND EPITHELIUM.

It may possibly be considered a new or unusual thing to devote a chapter to a consideration of the construction and function of the mucous membrane and epithelium; but *pro novo homine medico*, it will doubtless be well to call attention to the fact that, physiologically, there is no organ of such fundamental, vital importance to the nourishment of the human body as a healthy mucous membrane; for there are very few pathological conditions in which it is not primarily, or becomes secondarily, involved; and, when defective, it exerts an important influence in retarding the proper nourishment of certain parts, or of the entire system. If more attention were paid to the condition of the mucous membrane, especially of the lower portion of the œsophagus, and of the stomach, and to their abnormal acidities, many cases of starvation would be prevented, and death certificates signed "phthisis," or "consumption," would be considerably reduced in number. That abnormal acidity is always the cause of digestive disturbance must not be inferred, as many vital pathological conditions of near or remote parts may also produce digestive derangement; but reference is made to abnormal acidity simply to indicate the necessity of beginning observations at the foundation—the mucous membrane. In nine cases out of ten the trouble will be found there; for the mucous membrane of the lower portion of the œsophagus, from the diaphragmatic œsophageal opening to the cardiac orifice, is the first severely affected by the ingesta, be that solid or fluid, hot or cold, exciting, depressing, irritating, or inflaming, distending or contracting, wholesome or harmful. That these influences may easily produce an abnormal secretion of acid for the gastric juice must be seen at a glance. Again, there can be no doubt that *ptomaines* (alkaloids derived from dead animal tissues or decomposing proteids) exert their influence to a much wider extent in pathological conditions than is at present suspected. In no tissue of the body are ptomaines developed so favorably as in the mucous membrane, as it is very delicate, easily degenerates (though it rapidly repairs, generally), and is in direct contact with the atmosphere and its animalcula, which find in the ptomaine, or defective or dead parts of the mucous membrane, a most favorable soil to multiply, and especially to develop *saprophytes* or *pathogenytes*. When a ptomaine is once developed, its animalculæ product is easily absorbed and carried by the lymph and blood to a distant organ, which then suffers. The lung is continually active in extending and contracting, and is very delicate in structure, which no doubt accounts for its being affected in by far the greater number of all diseases; and it is not difficult to comprehend that the primary cause of pulmonary tuberculosis is found in a mucous membrane, be that the Schneiderian, respiratory, genito-urinary, or alimentary—especially the latter, from whence the lungs may become subsequently affected.

Mucous Membrane.—In structure the mucous membranes are soft, spongy, and velvety; well supplied with blood-vessels, lymphatics, nerves, follicles, and minute glands imbedded in their fibrous and submucous tissues. The outer layer of the mucous membrane has a free surface formed of epithelial cells, resting on a thin, transparent, homogeneous layer, termed basement membrane, beneath which is a *stratum*, or layer, of fibrous tissue adhering to the submucous structure, and well interwoven with vascular tissue (Figs. 8 and 13). They include the linings of all internal passages that communicate externally with the skin, such as at the nose, eyes, ears, alimentary, respiratory, and genito-urinary; but never form completely closed sacs as do the serous membranes. The mucous membranes are also very sensitive and irritable. Their cells secrete *mucus*, consisting of mucosin (mucin), water, and alkaline salts, such as potassium, soda, and lime phosphates, keeping the membrane moist so as to better perform its functions. **Mucus** is peculiarly viscid, translucent, thickly glutinous, differing somewhat in consistency in different parts of the mucous membrane, the surface of which it covers. Normally, it is of thicker consistency than the fluid of serous membranes. According to location, the surface of the mucous membrane is either in the form of villi, projecting papillæ, or depressions; the latter being in the form of ducts of either follicles or glands (Figs. 8 and 13). In the stomach, uterus, and bladder the submucous tissue forms a distensible layer, so as to permit of considerable extension without injury. In contraction of these organs the mucous membrane is thrown into folds.

The mucous membrane of the alimentary canal commences at

the lips and ends at the anus; its glands and follicles furnishing the
true digestive secretions for the reduction of food, while the villi
absorb the nutrient materials thus reduced to a liquid.   The salivary glands furnish saliva; the lower portion of the œsophagus the
acid, and the gastric follicles the pepsin of the gastric juice; the
glands and follicles of the small intestine the intestinal secretions;
while the membrane of the large intestine also absorbs fluid substances, and constitutes an outlet, that emits not only the indigestible residue of food, but its numerous minute excretory follicles,
lined with epithelial cells, separate products from the blood that
must be excreted to prevent disease.   (*Vide* chapters on Digestion,
Secretion and Excretion.)

The mucous membrane of the nostrils extends through the lachrymal duct, forming a junction with that of the eye.   In the vault
of the pharynx it continues the lining of the Eustachian tube and
wall of the cavity of the tympanum (middle ear); while in the
pharynx it continues as the pharyngeal mucous membrane to the
alimentary tract.

Through the alimentary and respiratory mucous membrane the
elementary nutritious materials are absorbed for the body.   The
cells of the secreting follicles and glands of the alimentary mucous
membrane produce the various digestive fluids; and, that the nutritive materials may be of benefit to the system, they are absorbed
by absorbent vessels in the membrane and conveyed into the blood
circulation.

The respiratory mucous membrane lines the larynx and trachea,
and extends to the bronchial tubules.   Each tubule is about 1-60th
of an inch in diameter.   From this situation on, the membrane is
very delicate, and continues to the pulmonary alveoli (air-cells),
and serves for the absorption of oxygen and the exhalation of carbonic acid gas, as well as for keeping the air passages moist with a
layer of mucus, protecting the membrane from irritation. If congestion exists, as in a certain stage of bronchitis, this mucus secretion
is at first diminished and afterward increased.   It is also increased
by administering either apomorphine, ipecac, emetine, or pilocarpine, while it is diminished by belladonna (atropia) and opium
(morphia).   In expelling copious secretions from the bronchial
tubes, expectorants may operate in one of two ways—either by
local stimuli to the bronchial or gastro-œsophageal mucous membrane, forwarding an increased afferent stimulus to the medulla
oblongata, which sends a reflex impulse correspondingly through

the phrenic and intercostal nerves to the respiratory muscles, effecting forcible expectoration (mechanical), such as results from administering ipecacuanha or lobelia; or by means which change the
secretion at the same time, as apomorphine, pilocarpine, or emetine. This change of secretion is materially assisted by warm drinks
or moist inhalations. (*Vide* chapter on Respiration.)

The genito-urinary mucous membrane begins at the genito-urinary orifices, and lines the urethra, bladder, ureters, pelvis of the
kidneys, vagina, uterus, and Fallopian tubes.

The moment any mucous membrane becomes much congested
and inflamed, the intercellular epithelial spaces become blocked
and the mucus-secreting cells beneath restrained in activity, while
the epithelial cells lining the membrane and ducts are much more
active, but deprived of their protector (the mucus); consequently,
their secretion becomes abnormally increased and acid. The epithelial cells and glands of the lower portion of the œsophagus are
very numerous, and placed in ring-like arrangements above and
around the inner surface of the cardiac orifice, which, being habit-
ually closed, retains the food more or less in that portion; so that
when food is introduced, especially of a solid, irritating nature, or
difficult of digestion, these cells and glands become very active; the
congestion of the membrane is abnormally severe, and the acid secretion is increased. Every one must have experienced, after the
introduction of such food, the excessive sourness of the digestive
process. This process of vital tissue acid secretion (a form of lactic
acid, and the only acid normally formed primarily,) also applies
to the stomach, where the lactic acid becomes generally changed into
hydrochloric acid. The acid secretion of the lower portion of the
œsophagus, and of the stomach, though in the latter in a less degree,
takes place normally when food of more or less difficult digestion
is taken; but when it occurs in the small intestine it is entirely abnormal, and gives rise to a pathological condition.

Next in order, as a suffering organ, is the stomach. When its
mucous membrane is continually irritated, the mucus secretion
becomes diminished; consequently, its natural protector, the alkaline mucus, is wanting, and the membrane softens. The acid secreted has then full play, not only on the ingesta, but also on the cells
of the membrane that produced the acid. Now, any alcoholic
drink, sugar, or food prepared from pure starch, which becomes
changed during digestion into sugar, or any other solvent or irritant to said membrane, increases the acid secretion and softening;

and, if continued, there can be no other termination than indigestion, diminished absorption, and starvation of the body.

That the stomach *normally* secretes very little or no acid, was observed in 1834 by Eberle—"that by maceration of gastric mucous membrane in acidulated liquid, an artificial gastric juice is obtained." This shows that the pepsin is in the gastric membrane (some of the pepsin being retained, possibly, in the gastric follicles), *but the acid has to be added.* Physiologically, the function of the lower portion of the œsophagus is important, as it secretes normally the acid for the gastric juice; but when the acid secretion is abnormally increased, it exerts a powerful influence in reference to indigestion, diarrhœa, anæmia, inanition, and consumption. (*Vide* chapters on Secretion, and Digestion.)

Endothelium consists of one or more thin layers of cells, destitute of blood-vessels and nerves. It forms the lining of internal *closed* cavities. Its free surface (examples of which are found in the membrane of the cavities of the brain, of the central canal of the spinal cord, of the interior of lymphatics, blood-vessels, serous, cardiac and synovial cavities, and the membranous labyrinth of the internal ear) is not exposed directly to the air. The endothelium of these structures consists of flattened, irregular squamous cells, adjusted to each other closely, and so presenting a smooth surface.

Epithelium consists of one or more thin layers of cells cemented together by a clear albuminous intercellular substance. It covers the entire free surface of the body and internal *open* passages, and such organs as the liver, kidneys, lips, etc. A free surface is a structure not blended with, nor attached to, adjacent structures, but is free, or separable from them without the necessity of dissection. Epithelium covers not only the entire external surface of the body, as the epidermis of the skin and its appendages (the nails and hair); but also the lips, membrane of the nose, and mouth, and continues over the entire mucous membranes of the internal *open* passages, such as the respiratory, alimentary, genito-urinary, the lining membranes of ducts of secreting follicles and glands, the membrane of the anus, external auditory canal, and of the tympanum, or middle ear.

Epithelial Cells are of various shapes, and may be classified into five varieties—ciliated, columnar, squamous, glandular, and transitional. Each class of cells has a special function to perform in the economy, though all *serve one purpose* in common; viz., to protect the surface of the respective membranes which they cover. This function is assisted by secreting cells lying beneath and partly between the epithelial cells, and secretes more or less moisture over the epithelium. Thus, the cells underneath seem to appreciate the kindness of their protectors, so that when the epithelium becomes irritated or inflamed, especially that of the mucous membrane, the mucus-secreting cells immediately increase their activity by furnishing an increase of mucus, which passes by exosmosis through the albuminous intercellular epithelial substance to the surface of the membrane for its protection. This view seems more tenable than that which holds the mucus secretion to be a gradual dissolution of the outermost epithelial cells. The principal constituent of mucus is mucosin, one-eighth of which is nitrogen. By the mucosin nitrogen is considerably eliminated from the blood, which may explain the anæmic condition of persons having a defective mucous membrane or insufficiency of mucus secretion, nitrogen being, in this case, insufficiently eliminated.

The **Ciliated** variety (A, Fig. 3) consists of columnar cells (1) internally pointed, each cell containing a nucleus (3). On the sur-

Fig. 3.

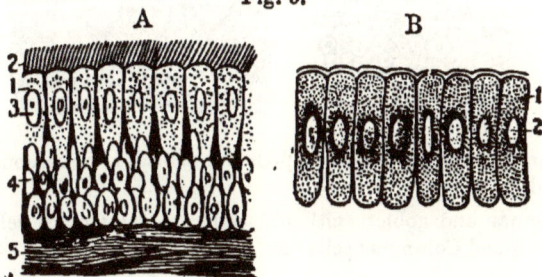

A—1, Ciliated columnar epithelial cells.   2, Cilia.   3, Nucleus.
4, Young cells. 5, Elastic basement membranous and fibrous tissue.
B—Columnar epithelial cells. 1, Club shaped. 2, Its nucleus.

face of the mucous membrane these cells have projecting cilia (2) of about 1-3000th of an inch in length, which vibrate about 700 times per minute in the fluid which moistens the surface of the membrane, thereby propelling the mucus or other minute particles towards the orifice of the cavity or tube of the membrane which they line. These cells are found on the mucous membrane lining the nose (except the olfactory portion), lachrymal ducts, Eustachian tubes, tympanum, larynx, pharynx (excepting the vocal chords), trachea, bronchial tubes, uterus, Fallopian tubes, and vasa

efferentia.   Each cell may have from ten to thirty cilia, placed at
regular distances.

The **Columnar** variety (B, Fig. 3) consists of cylindrical columns or club-shaped nucleated cells (1), about 1-500th of an inch
in length, placed side by side perpendicularly.   They are found on
the mucous membrane lining the alimentary canal from the diaphragmatic œsophageal opening to the anus, secreting follicles, and
ducts of secreting glands, gall-bladder, and the urethra.   Sometimes,
near the surface of the mucous membrane, the upper end of these
cells are distended, forming the **Goblet Cells** (A, A, Fig. 4).

The **Squamous** variety of cells (1, Fig. 4) is a flat thin plate with
an oval nucleus, differing in shape and size in different localities.
These cells are found on the epidermis, hair, nails, internal lining
of the arteries, veins, capillaries, lymphatic vessels, acini of the
lungs, aqueous chambers of the eye, uriniferous tubules of the kidneys, and on the serous membranes (pleura, pericardium, and peritoneum).   The **Tessellated** epithelium consists of nucleated flat-

Fig. 4.

C, Squamous and Tessellated epithelial cells.   1, Squamous.   2,
Tessellated.

D, Columnar and goblet epithelial cells.   A, A, Goblet cells are
simply distended Columnar cells (B).

tened scales from 1-1000th to 1-600th of an inch in diameter (2, Fig.
4).   They are like the squamous, but considerably larger, and
scattered among them.   They are found in nearly all localities
mentioned with the squamous; and also on the mucous lining of
the mouth, pharynx, œsophagus, vestibular entrance to the nose,
conjunctiva, entrances to the urethra and vagina, and the greater
portion of the epidermis.

The **Glandular** epithelial layer (E, Fig. 5) consists of nucleated cells.   They are found in the acini of the various secreting
glands, and in the convoluted uriniferous tubes of the kidneys.
These cells separate from the blood the materials out of which

they produce or secrete the different fluids, follicular and gland-ular juices.

The **Transitional** epithelial cells (H, Fig. 5) occur in many different shapes, sizes, and localities, all having a flattened surface, and some a rounded extremity. These cells are found on the mu-cous membrane of the bladder, pelvis of the kidney, and ureter; they are often intermixed with the columnar epithelial cells, as in the larynx and pharynx.

All varieties of epithelial cells are, as a rule, arranged in layers, or strata. The cells next to the surface have a tendency to be shed, and their place is then taken by the cells of the deeper lay-ers, which become modified in form as they approach the surface. The cells of glands, also, are constantly cast off; yet each gland maintains its form and proper composition, and for every cell

Fig. 5.

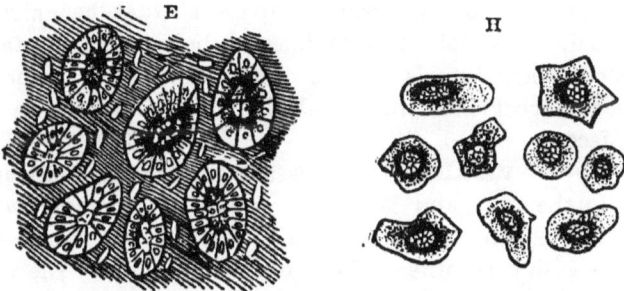

E, Glandular epithelial cells.   H, Transitional epithelial cells.

thrown off a new one is produced. The epidermis furnishes a good example, the surface cells thereof be constantly thrown off by those newly formed beneath. (*Vide* chapters on Secretion, Absorption, Digestion, and the Skin.)

CHAPTER IV.

DIGESTION.

The **Digestive Organs** include the mouth, stomach, and small intestines; the **Digestive Apparatus,** however, may be said to include the mouth, œsophagus, stomach, small intestines, liver, and pancreas. Each one of these organs secretes a particular digestive juice differing from the others (the œsophagus secretes a liquid acid), each secretion performing its particular work in the process of digestion.

Digestion is the process of reducing the materials of the food to a semi-fluid form, fitting it for absorption into the circulation and for assimilation. It is partly a mechanical process (mastication), and partly a chemical process (solution and transformation by the juices). **Food** is composed of organic and inorganic substances, and to be of use in the economy must be assimilated. Animal life, unlike vegetable, requires material that has already been organized, and which must be digested before it can be taken up by the absorbent vessels. Some of the digested, or liquefied, food is then carried into the circulation; the remainder is discharged as feces.

The **Alimentary Canal** is the passage through the body from the mouth to the anus, and is divided into several compartments by narrow constrictions (guards), through which the compartments communicate with each other. The mouth is guarded by the valve of the isthmus of the fauces; the œsophagus is guarded at the cardiac and the stomach at the pyloric orifice. Then follows the small intestine, about twenty-five feet in length in the adult. Anatomically and physiologically, the small intestine is subdivided into three parts; *viz.,* the *duodenum, jejunum,* and *ileum.* In the duodenum are the orifices of the biliary and pancreatic ducts; at the end of the ileum is the guard termed ileo-cæcal valve; then follows the large intestine, which is guarded by the sigmoid flexure; then passes on and terminates at the anus, guarded by a double sphincter muscle. The ileo-cæcal valve allows fecal matter, or that which has not been appropriated by the system, to pass from the small into the large intestine; so that in the normal state no regurgitation of the ingesta from the large to the small intestine occurs. The large intestine (colon) is about five feet in length in the adult. Its contents differ much from those of the small intestine, especially in being more solid. In passing towards the exit the contents become more solid, in consequence of the absorption of the fluid by the mucous membrane. The *vermiform appendix* (8, Fig. 6), below the ileo-cœcal valve, may possibly serve as a large secreting follicle to lubricate the acute curve at the valve with a large quantity of an alkaline secretion. When the alkaline secretion of the ileum is deficient and acidity occurs—caused especially by irritating food, and certain kinds of fruit—great pain results; but the alkaline secretion of this appendix may subdue the irritation and pain, by neutralizing the acidity.

The **Stomach** (3, Fig. 6) consists of four different tissue layers— mucous, submucous (fibrous), muscular, and serous (Fig. 13). The serous is the most external, the mucous the most internal, the other two being situated between them. The wall of the stomach averages about a line, or 1-12th of an inch, in thickness. Its muscular layer is about one-half of the thickness of the wall, and its fibres are arranged in three sets—the *longitudinal*, externally; the *oblique*, internally; and the *circular*, between. When the stomach is distended to 12 or 15 inches laterally and about five antero-posteriorly, it has a capacity of about four or five pints. The stomach is supplied with blood-vessels, lymphatics, nerves, and secreting follicles; the latter consisting of tubes set

Fig. 6.

Diagram of Alimentary Canal from Œsophagus to Anus.

1, Œsophagus. 2, Diaphragm. 3, Stomach. 4, Duodenum. 5, Small intestine. 6, Ileo-cœcal valve. 7, Cæcum. 8, Vermiform appendix. 9, Ascending colon. 10, Transverse colon. 11, Descending colon. 12, Sigmoid flexure. 13, Rectum. 14, Diaphragm attached to lower part of chest-wall. 15, Spleen.

closely together transversely in the mucous membrane. The follicles (A, A, Fig. 13) are from about 1-60th to 1-30th of an inch in length, and terminate in blind extremities in the wall of the stomach. The mouths of the ducts of these follicles are from 1-500th to 1-300th of an inch in diameter, and open on the surface of the mucous membrane, in the interspaces between its projecting folds. The wall of each follicle is provided with columnar epithelial secreting cells (D), ovoid in shape, each about 1-1200th of an inch in diameter. The surface of the mucous membrane is covered with a layer of transparent columnar epithelial cells (E), which secrete a thick, tenacious, alkaline mucus, for the protection of the mucous membrane of the stomach. The cells secreting the mucus of the membrane are active even during fasting; while the cells of the gastric follicles secreting the pepsin are active only when food or a foreign substance enters the stomach. The stimulus for the secretion of pepsin and for the peristaltic activity of the stomach is due to the sensation of touch caused by the presence of the ingesta. As all external sensations (as will be noticed in the chapter on the Senses) require for their activity a combination of three distinct but associated organs—viz., the peripheral organ (or cells) of sensation, a nerve fibre or fibres, and a nervous centre (or cells)—so it is in regard to the sensation of touch in the stomach. The presence of ingesta causes the sensation in the stomach, inducing a powerful stimulus, which is carried through afferent fibres of the pneumogastric nerve to the medulla oblongata, which reflects the stimulus by sending a vaso-motor impulse through the efferent fibres of the splanchnic nerve, passing through the gastric plexus and entering the wall of the stomach. The efferent (vaso-motor) fibres induce dilatation of the capillary vessels, so that the secreting cells of the gastric follicles in the wall of the stomach receive abundant material for the secretion of the pepsin of the gastric juice. The acid of the gastric juice is not secreted by the cells of the gastric follicles. (*Vide* Gastric Juice.) The efferent impulse to the cells of the muscular fibres of the stomach causes them to become very active, at the same time inducing the peristaltic action—effected by the peculiar arrangement of the *longitudinal*, *oblique*, and *circular* layers of the muscular fibres. The peristaltic action of the stomach is essential to digestion, bringing all the food in contact with the gastric juice, causing intimate mixture and solution of the food.

Digestion of Food is effected by mastication and solution;

the latter being brought about partly by the saliva and partly by
the digestive juices in the alimentary canal. **Mastication** by the
mouth is essential before the food is properly fit for the dissolving
changes during the digestive process, not only on account of the
physically fine reduction and solution of the materials, but also
on account of the action of the saliva, termed *insalivation*, which
is of a chemical nature, acting on the insoluble starch of food and
converting it into soluble *glucose* (grape sugar). The active principle effecting this chemical change is the *ptyalin* of the saliva.
Raw starch is indigestible by the human stomach, and often passes
unchanged from the bowels; but when thoroughly hydrated it is
easily acted on, and transformed by the saliva and pancreatic juice
into glucose. It is for this reason, also, that starchy vegetables
require thorough solution by cooking to render them digestible.
Sago, tapioca, and arrow-root are composed of starch almost entirely;
wheat contains 70 per cent, and rice 90. Peas, beans, chestnuts,
acorns, rye, oats, and potatoes also contain considerable starch.
Normal gastric juice does not change starch into glucose; neither
can the saliva accomplish this change in the stomach, on account
of the *acid* of the juice; but the action of moisture, heat, and acid
dissolves the starch and separates the *dextrin*, which is isomeric
with glucose, and is like gum arabic, sticky and adhesive. During
digestion the dextrin is dissolved into *dextrose*, which is similar to
glucose in that it polarizes light to the right, but it has not the significance of glucose in digestion nor in nutritive properties. The
movements of the stomach during digestion are like the process of
kneading. The food is moved toward the left of the stomach into its
great pouch, then downward, and along its greater curvature to the
pyloric portion, and finally carried to and fro to every part of the
organ and mixed with the gastric juice. The product of the solution and disintegration of the food substances in the stomach is
termed **chyme**. The normal secretion of the mucous membrane
of the empty stomach is alkaline. (*Vide* chapter on Mucous Membrane.) The stimulus of food promotes the activity of the cells
of the follicles and the consequent secretion of the normal pepsin
of the digestive gastric juice. (*Vide* Fig. 13.) If the action be too
prolonged, however, the cells become exhausted, and any excess of
food must remain undigested and pass into the intestine in a crude
state, causing irritation and acidity until expelled. In disease, the
normal stimulus by the food and the normal functions of the secreting cells are more or less impaired, consequently there is no

4

desire for food. If food is taken then, it does not stimulate the cells to the secretion of gastric juice, but remains undigested, adding irritation to the other morbid conditions. The secretion of gastric juice is influenced by, though not dependent on, nervous agency. Strong mental disturbance, unusual fatigue, also febrile conditions, interfere with the digestive process—possibly by weakening the stimulation on the gastric membrane and the reflex action of the nervous centre cells; hence, there are no, or only impaired, vaso-motor impulses. Gastric digestion consists in a dissolution of the food by the gastric juice, the action of this juice being assisted by the peristaltic movements of the stomach. A temperature of from 98.5° to 100° F. is required for the perfect action of the juice; hence, cold and iced substances taken to excess depress the temperature of the stomach and are injurious. Moderate exercise before and rest after a meal facilitates digestion. As soon as the gastric contents pass into the duodenum, the action of the pepsin and the acidity of the chyme suddenly cease on account of the action of the bile pre-cipitating the pepsin, and the alkaline pancreatic and intestinal juices neutralizing the acid. The bile and the alkaline secretions then further act upon, dissolve, and disintegrate the peptones and other ingredients of the chyme, transforming it into **chyle**. In that manner the peptones become real albuminose, fibrinose, and globulose, which, when absorbed by the lacteals, become albumen, fibrin, and other essential material of the circulating blood. It is by the digestive processes that the food substances are dissolved and the chemical atoms and molecules liberated and enabled to form new compounds in the economy. The nutritive substances enter the blood-circulation, and the cells of the secreting follicles and glands secrete from the blood the many different active principles: the *albuminoids*, or so-called *proteids*, of solids and fluids—such as hæmoglobin, casein, ptyalin, pepsin, pancreatin, mucosin, myo-sin, melanin, collagen, chondrin, elasticin, and keratin. Again, the liberated atoms and molecules of properly digested food furnish also the material for the construction and formation of the many differ-ent tissues. These facts indicate that it is of vital importance to the economy that the food taken be always properly masticated and digested.

Food in its passage through the alimentary canal is acted upon by at least five different secretions; viz., saliva, gastric juice, bile, pancreatic and intestinal juices, which produce successively not only simple solutions, but also more or less transformations of the food into new substances. The products are then absorbed, some by the capillary veins, others by the lacteals of the small intestine, and become mingled with the blood; the indigestible substances pass through the cæcal valve into the large intestine, and are finally expelled with the feces. (*Vide* chapter on Absorbent and Lymphatic System.)

Gastric Juice.—The cells of the mucous membrane of the lower portion of the œsophagus and at the region of the cardiac orifice of the stomach secrete principally the *acid*, which is a result of the irritation of the part. This acid may be called a *vital tissue acid*. It is similar in composition to lactic acid. Through the sodium chloride of food, heat, and the moisture of the stomach, this acid is generally changed into hydrochloric acid; but, when milk or starchy food only is taken, the acid remains lactic acid, instead of being transformed into hydrochloric. The cells in the walls of the follicles of the mucous membrane of the stomach secrete from the blood of the arterial capillaries an albuminoid (proteid) termed **pepsin**. This is the characteristic active principle of the gastric juice. It requires for its action, however, the presence of a free acid, either hydrochloric or lactic—the pepsin in the stomach cannot act upon animal food without an acid reaction. An acid, therefore, is first secreted as soon as the food passes through the lower portion of the œsophagus. The acid-secreting cells are very active, so that when the last portion of the meal has passed into the stomach the lower part of the œsophagus is still in a state of irritation and in an acid condition. This acidity is gradually neutralized during digestion by the slightly alkaline saliva swallowed; consequently, to a person with deficient saliva or delicate mucous membrane of the œsophagus, expectorating the saliva soon after a meal is very injurious, as the acidity of the œsophagus is not relieved. Physiologically, it would be out of question to treat such a case for gastric indigestion or dyspepsia, and make the stomach suffer when it is not at fault. Of course, the alkalinity and the acidity can only manifest their proper action when the œosphageal and gastric secretions are in normal condition. When the normal acid of the gastric juice is deficient or suppressed, the administration of properly diluted hydrochloric acid will prove beneficial. On the other hand, when the acid secretion is normal, or excessive, then the administration of an acid diminishes gastric digestion. Failure of sufficient secretion of acid lies generally in an abnormal thickening of the mucous membrane of the lower portion of the œsophagus, lessening the irritation by the food substances on this organ. The eating or drinking of irritating, fermentive, or indigestible liquid or food generally causes increased secretion of acïd. It would be very difficult to succeed in treatment where there is an impaired mucous membrane and increased acidity at the lower portion of the œsophagus if any alcoholic fluid, sugar, or food prepared from pure starch (which during digestion is transformed into sugar) is allowed to be taken. (*Vide* Stomach, the Digestion of Food, and the chapter on Mucous Membrane.)

Pepsin is the albuminoid constituent of the gastric juice, converting food proteids (albuminous or nitrogenous substances) into *peptones*. Its action seems to be limited to the reduction and complete solution (digestion) of albuminous food, such as eggs, meat, cheese, gelatin, and vegetable glutinous foods, embracing the blood-forming portions of the vegetable and all the nitrogenous elements, such as the albumen, fibrin, globulin, and casein of animal food (except the fat—non-nitrogenous), and converting them into purified compounds termed **peptones;** which, in composition, resemble the substances from which they are derived. The *conversion* of the food substances is effected by the power of the pepsin, and their *solution* by the acid of the gastric juice; they are disintegrated, softened, and their different atoms and molecules liberated, each class of atoms and molecules coming into closer relation, forming albumen-peptone, casein-peptone, fibrin-peptone, and others. As it is principally the pepsin of the gastric juice which effects these changes, the products are called *peptones*. It is generally conceded that this separation and formation of compounds, or peptones, takes place in the stomach; though, as previously stated, digestion is not completed in the stomach, but continues in the intestines. Here these peptones, coming in contact with the intestinal juices, are more finely broken up; for instance, the albumen-peptone, coming in contact with the pancreatic juice in the duodenum, is purified and becomes that true albuminose capable of absorption by the lacteals of the mucous membrane, after which it becomes a part of the serum-albumen of the circulating blood. Other peptones are similarly acted upon and absorbed. A natural peptone can only be produced by the action of the pepsin and the acid of the gastric juice in the stomach, and not by pancreatic juice, or any other organ or substance. Nature, in all its acts, is simple; therefore, it is only by misconstruing natural acts or their results that we go further from the comprehension of its laws. Physiological chemistry, considers the peptones of milk, casein, albumen, and fibrin separately. Analysis proves them similar in composition to the substances they are derived from; therefore, the peptones constitute an intermediate stage between the food just entering the stomach and the substances absorbed, as dough constitutes the intermediate stage between flour and bread. The albumen-peptone lies between the albumen of the food and the albuminose ready to be absorbed; the fibrin-peptone, between the fibrin of food and fibrinose; likewise, the other peptones.

Lehman gave the name **peptone** to the dissolved proteids of food in the stomach, the solution being effected by the action of the pepsin and the hydrochloric acid. Landois tells us that "peptones represent the highest degree of hydration of the proteids." According to the *Schweizerische Wochenschrift fur Pharmacie*, "the best test for distinguishing albumen from peptone is xanthogenate of potassium, which precipitates albumen in a neutral solution, only after the addition of an acid. Peptones are precipitated at once, as they already contain free acid." The foregoing substantiates the assertion that peptones hold an intermediate state between the food entering the stomach and the substances dissolved, purified, and ready for absorption in the intestines; for the alkaline condition of the intestinal juices neutralize the acid, while the bile acts upon the pepsin. Again, the peptones undergo other modifications. In short, the peptones are not absorbed as such, but form the materials which, through the action of said secreted juices, become true albuminose, fibrinose, globulose, which enter the circulation, and these again undergo further changes, as will be noticed in the following pages.

It may be well not to lose sight of the difference between "albumen," "albuminous," "albuminose," and "albuminoid"; as one compound cannot be the other; neither can one, in the processes of life, be used in place of the other. The first three mentioned have been partially considered already, but the following may further illustrate them, and especially the last—the albuminoid. The change of albuminous substances into albuminose, fibrinose, globulose, and other solutions, as long as they remain in the alimentary canal, must be considered simply as a physical change (in this case a fine solution), since a *physical change* does not cause the loss of the identity of a substance. This is illustrated by the fibre of linen, which is identical with the fibre of the mature flax-plant; or by dissolved sugar in water which is identical with sugar in solid form. When albuminous substances are transformed in the stomach into albuminose, fibrinose, or globulose, and have entered the circulation, then, by the protoplasm of the tissue cells (this includes blood-cells, or corpuscles), chemical changes and combinations take place, producing *albuminoids* (proteids). The protoplasm of blood-cells (corpuscles) produces the albuminoid hæmoglobin; the protoplasm of the cells of the secreting organs, the albuminoids of digestive juices, such as ptyalin, pepsin, pancreatin, etc.; and the protoplasm of the cells of the solid tissues the albuminoids, myosin, chondrin, etc.; the blood furnishing the material for the secretions and the transformations.

We have now arrived at a distinctive point. There are two classes of albuminoids—*juices* and *solids*—which are very distinct in their properties, but similar in composition, and act differently. Those of the juices act in dissolving and transforming, but exhaust themselves during the process; such as the ptyalin of the saliva, the pepsin, pancreatin, etc. Those of the solid tissues are solidifying, not transforming, and maintain themselves; such as the myosin of muscles, chondrin of cartilages, and others. The animal solid albuminoids have some analogy to the alkaloids of the plants, such as quinia of cinchona bark, or strychnia of nux vomica, excepting that the latter contain no sulphur. An **albuminoid** is a *nitrogenous compound*, also called **proteid** (holding the first place); and is composed of C, O, N, H, and S. In elementary composition the various albuminoids are alike, but differ in the number of atoms composing each. To illustrate: one albuminoid may contain more atoms of carbon, another of nitrogen, another of hydrogen, oxygen, or sulphur, and so on. Here we notice that the term "albuminoid" may indicate its derivation; but, as it is the product of a chemical change, it is not identical with the substance from which it is derived. This is illustrated in invisible coal-gas, which is obtained from visible solid coal. *Chemical* action separates the atoms, which then unite with others; the molecules and substance formed lose their identity. Now, the food and certain fluids entering the body and containing the essential elements, may properly be called *protein bodies*, or substances out of which the numerous *protein compounds* (albuminoids) are formed in the economy; such as the ptyalin, pepsin, pancreatin, myosin, and others. The average albuminoid compound is, chemically considered, very complicated, containing about 54 per cent of carbon, 22 of oxygen, 16 of nitrogen, 7 of hydrogen, and 1 of sulphur. The various proteids formed in the body from the food protein compounds differ only in their atomic proportions, and exist naturally in the body, either in a semi-solid or in a solid condition, some of them being easily changed into a more or less insoluble state by a process termed *coagulation*. This coagulation is brought about in fibrin, for instance, by simply allowing the fluid containing its elements to escape from a vessel of the body; other substances containing an albuminoid, coagulate either by boiling, by contact with mineral acids, or with numerous other substances (such as the pepsin in contact with bile). The pepsin of the gastric juice is precipitated by contact with bile; hence, if bile enters the stomach, it either disturbs the digestion or causes vomiting.

Indigestion is often the result of an imperfect secretion of acid or of pepsin; weakness of the mucous membrane of the stomach; or of fermentation, due to micro-organisms; either one of which changes the process of digestion into an abnormally increased acid condition, gas being evolved at the same time. All substances that contain much tannin, when taken into the stomach, also retard the digestive process. The rapidity with which watery solutions—for instance, of iodide of potassium, the alkaline carbonates, lactates, citrates, and hepatic remedies—pass into the blood, thence into the liver, urine, and saliva proves that the capillary veins of the stomach take them up; and, as they are tributary to the portal vein, the absorbed matters pass directly to the liver and stimulate it to increased secretion.

The mucous membrane of the alimentary canal is not the only channel through which nutritive or other substances may be introduced into the system; as water, and substances dissolved in it, may get into the circulation by absorption through the skin. When no food can be taken into the stomach for some time, life may be sustained by enema, or by bathing in milk, bouillon, etc. In this case, the lymphatics of the skin absorb the fluid food and carry it into the blood circulation. It may be here observed that a chemical change of such nutritious substances before absorption is not necessary; but a thorough solution simply is required. The lymphatics (lacteals) of the intestine have the advantage of position to absorb the nutritious substances more thoroughly than the lymphatics of the skin. Food in the alimentary canal is, with respect to the bodily tissues, yet outside the confines of the body, as a fly inclosed by the leaf of an insectivorous plant is yet outside the tissues of the plant itself. The cells of the lacteals are dis-

posed to busy themselves during the digestive process with absorbing the more intricate materials, such as albuminose, fibrinose, and
fatty emulsion; while the capillary veins of the intestinal wall
absorb not only venous blood, but also glycogen, or saccharine matter, and discharge them into the portal vein.

That the stomach does not dissolve or digest *itself* may find an
explanation in one or more of the following four circumstances:
1. No digestive pepsin is secreted when the stomach is empty. On
administering pepsin or some other artificial digestive substance
when the stomach is empty, the mucous membrane is attacked
(provided a free acid is present), and pain results; this may also
happen when the dose of pepsin is too large, with little food in the
stomach. 2. The alkalinity of the blood, which circulates copiously
in the layers of the wall of the stomach, counteracts an attack
by the normal acid of the gastric juice. 3. Vital activity opposes
the digestion of the stomach by itself, so that the chemical force
of the pepsin and acidity of the juice are expended on the non-living
food matter. 4. The indestructible transparent columnar epithelial cells of the gastric mucous membrane afford a natural protection against the self-digestion of the stomach. If, on the other
hand, at the completion of the gastric digestive process, the stomach
contains much pepsin and an abnormally increased amount of acid,
and the blood circulation is abruptly stopped, the gastric juice may
then act on the stomach, digest, and perhaps perforate it. This
may be of importance in a medico-legal sense.

Intestinal Digestion.—Chyme, or liquefied food, when passing from the stomach, is a mixture of the disintegrated or partly
digested food, some portions of which are completely liquefied,
others not. In the intestine the bile, pancreatic and intestinal
secretions act upon the chyme, thus continuing the digestion unfin-
ished by the stomach, transforming the chyme into **chyle,** which
is an opaque, milk-like fluid; and, when absorbed into the lacteals
it consists of oil-globules, with albuminous matter emulsified, and
fine, fatty granules, termed the *molecular basis of chyle.*

After the chyle has passed through the lymphatic glands of the
mesentery it also contains fibrin and white corpuscles (leucocytes),
derived from these glands. It is obvious, therefore, that chyle
gradually undergoes changes during its passage from the intestine
through the lacteals and lymphatic glands. When it reaches and
enters the thoracic duct it mingles and becomes identified with
the lymph.

The **Duodenum** is the immediate intestinal continuation from the stomach, the pyloric sphincter and orifice being the dividing line. It is from ten to twelve inches long. At a point about four inches from the stomach, the excretory duct (*ductus communis choledochus*) of the liver and gall-bladder enters and empties itself. Just below the orifice of this common bile-duct is the orifice of the larger pancreatic duct, and a short distance below this is the orifice of the smaller pancreatic duct. The jejunum and ileum have no distinct line of demarkation between them. The duodenum, jejunum, and ileum together are about five times as long as the human body. The jejunum is so called from the fact that it is generally found empty.

The **Wall of the Intestine** (Fig. 8) is composed of four tissue layers—a serous externally, a mucous membrane internally, and between these two a submucous connective tissue, and a muscular-fibre layer. The muscular layer is arranged in diagonal, longitudinal, and transverse layers of muscular fibres; their alternate contraction and relaxation producing the so-called **peristaltic action**, which propels the intestinal contents forward. The presence of food in the intestines is the appropriate stimulus to excite the sensory cells of those parts. These cells communicate freely with the sympathetic centre-cells and those of the medulla oblongata. In this manner reflex actions are effected, producing the peristalsis, or alternate contraction and relaxation of the muscular fibres in the wall of the intestine. The mucous membrane of the small intestine is thickly covered with **villi**, which, at their surface, are lined with a layer of epithelial cells (A, A, Fig. 7; and B, Fig. 8). When the liquefied food (chyle) passes along, each villus is completely submerged in the chyle. The villi are evidently intended not only as absorbent agents, but also to increase the absorbent surface. Each villus is supplied with a set of blood and lacteal vessels, the latter discharging into lacteal follicles beneath. The villi are most highly developed in the duodenum, where the chyle is first formed. They are turgid, enlarged, and opaque during the process of digestion, but in prolonged fasting become shrunken from inactivity and diminished nutrition.

Fig. 7.

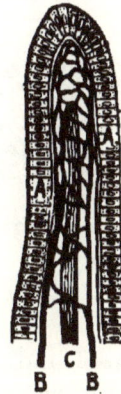

Diagram of a Villus.

A, A, Epithelial cells. B, B, An arterial capillary and a capillary vein. C, Lacteal vessel.

These villi are very active agents during the process of absorption of chyle in the small intestine, which has about 6000 square centimetres of surface, with from 15 to 18 millions of villi at least. In the layer beneath the villi of the mucous membrane of the small intestine are found the follicles of Lieberkühn (C, C, Fig. 8), uniform and parallel with one another, and opening between the villi on the surface of the mucous coat. These follicles are abundantly distributed in the small intestine, and possibly effect important changes in the digestive process. In the submucous coat (E) are small glandular clusters termed **Glands of Brunner** (D, D).

Fig. 8.

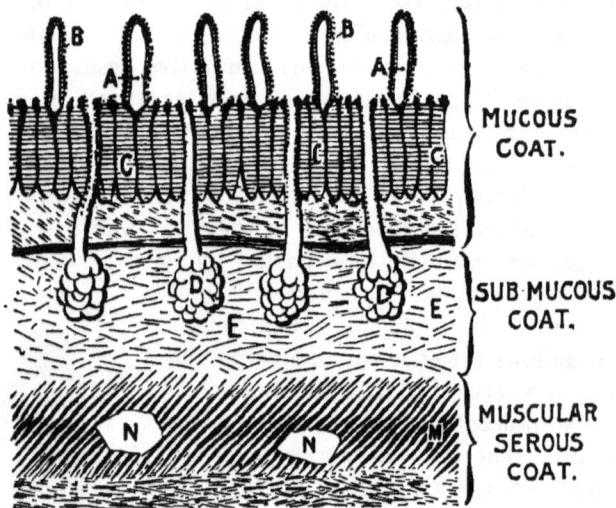

Diagram of the Wall of the Small Intestine.

A, A, Villi.   B, B, Epithelial layer.   C, C, Follicles of Lieberkühn. D, D, Brunner's glands.   E, Submucous coat.   M, Muscular coat, resting on the serous coat.   N, N, Ganglion cells.

They are found only in the duodenum, and are most abundant at its pyloric end. From the vesicles composing a cluster arise minute ducts, which coalesce and form a larger duct, through which the secretion of the gland passes into the duodenum.

The **Glandulæ Solitariæ**, or **Solitary Glands** (B. B, Fig. 9) are distributed throughout all parts of the small intestine. They are most numerous in the jejunum. These glands are simple, membranous, flask-shaped vesicles, the neck being even with the internal surface of the intestine, while the rounded base lies within the mucous membrane. These glands consist of granular particles, and have no openings. The manner in which they discharge their contents is not clearly understood. The solitary gland is never larger

Fig. 9.

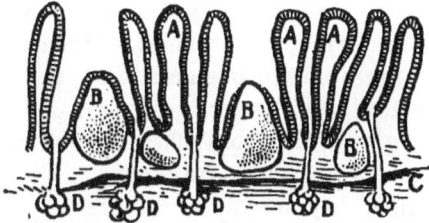

Diagram of Solitary Glands in the Mucous Membrane.

A, A, Villi. B, B, Solitary glands. C, Layer of longitudinal muscular fibres. D, D, Brunner's glands in the submucous layer.

than a mustard seed. Possibly, when attaining this its maximum size, it bursts and its contents escape. The mucous membrane of the small intestine presents also numerous transverse folds, termed **Valvulæ Conniventes** (Fig. 10). When a piece of this membrane is placed in water, the folds may be seen to move like the opening and closing of a fan. Each fold occupies three-fourths or more of the circumference of the mucous membrane and often projects more than half an inch into the intestine, supporting the villi, and checking somewhat the rapid motion of the intestinal contents.

Fig. 10.

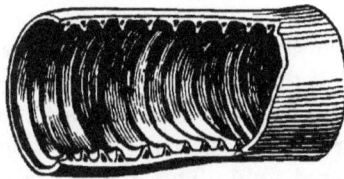

DIAGRAM SHOWING THE VALVULÆ CONNIVENTES OF THE SMALL INTESTINE.

Peyer's Glands are aggregations or clusters of solitary glands, forming oval patches. They are situated on the internal wall of the ileum, just opposite the attachment of the mesentery, and are smallest toward the jejunum, and largest near the cæcum. In the latter situation patches occur measuring from two to three inches in the long diameter. They vary in number from fifteen to twenty. Regarding the function of Peyer's glands, nothing certain is known; though, as they become enlarged during the digestive process, it is inferred that they in some way influence this process. It is also surmised that the peculiar odor of the feces may be due to the

presence of their secretion. In typhoid, or enteric, fever and in
phthisis, these glands become irritated, produce acidity, and generally ulcerate. This is probably the cause of the diarrhœa so common in these diseases.

The villi and the secreting follicles and glands in the wall of the
small intestine are nourished by blood from the capillaries of the
superior mesenteric artery. The branches of the arteries and veins
in the gastric and intestinal mucous membrane are surrounded by
a sheath of connective tissue, very strong along the arteries, but
delicate along the veins. These branches lay within the muscular
coat of the intestine in an oblique direction. On account of the
weakness of the veins, they are easily compressed by the contraction of the intestinal muscular coat during peristaltic contractions,
thereby impeding the return of the blood from the gastric and
intestinal mucous membranes, and increasing their congestion.
The congestion continues as long as the contraction lasts, and is
aggravated by frequent repetition. This condition of congestion
is, to the digestive process, of the greatest importance, as it favors
the longer retention of the blood in the walls of the stomach and
of the intestine, enabling them to furnish more material for follicular and glandular secretions. In health, the mere presence of food
in the stomach and intestines excites the contractions of their walls,
alternating with relaxations—peristalsis. These movements exert
a favorable influence upon secretion and absorption by favoring
not only the movement of the ingesta, but also by producing alternate compression and relaxation on the glands and follicles, thereby
favoring the flow of their secretion, and in the villi the flow of their
absorbed fluid. This demonstrates, also, that the congestion or periodic hyperæmia brought about during the peristaltic contractions
is really a physiological fact of vital importance. On the other
hand, severe irritation of the intestinal walls brings about acidity
and severe tonic contractions, resulting often in **diarrhœa.** In
this case the mucous membrane secretes acid instead of mucus, and
the natural secretions of the glands and follicles are diminished
and the digestive process seriously retarded or entirely suspended.
Should excessive irritation occur in the small intestine, the blood-
serum, with the albuminous substances, alkaline salts, and water
already absorbed, pass from the absorbent vessels back to the
intestinal canal, and are carried forward by strong peristaltic
movements so rapidly, that absorption in the large intestine cannot
keep pace with it, and diarrhœa ensues. In *cholera* the irritation

is most severe and fatal, because the blood thickens and as a result circulates sluggishly.

The mucous membrane of the large intestine is smooth and shining, and is destitute of villi; it is provided, however, with absorbing and excreting cells. In the walls of the stomach and intestines, the capillaries of the gastric and the superior and inferior mesenteric veins, respectively, are very active as absorbents.

Hunger and Thirst are peculiar sensations of desire manifested by the tissues of the body; hunger is referred to the stomach, thirst to the tongue and throat. When normal, these sensations indicate a want of the elements of food or drink in the tissues— hunger calls for food, thirst for water. When the tissues become satisfied, these sensations disappear The natural demand of the appetite affords the surest criterion of the quantity of food or drink required by the system.

CHAPTER V.

LIVER, BILE, AND PANCREATIC JUICE.

The **Liver** is the largest glandular organ in the body. It is situated beneath the diaphragm, in the epigastric and right hypochondriac regions, and is retained in position by five ligaments. Its size and weight are enormous when compared with other secreting glands. The weight of all the salivary glands together is only about three ounces, of the pancreas about two and a half ounces, while the liver weighs from three to four pounds. Its substance is divided into two large lobes, consisting internally of very small *lobules*, about 1-18th of an inch in diameter, and composed of clusters of *acini*. The interlobular connective tissue, or sheath, finally surrounding the lobule is called *Glisson's capsule*, being a continuation of the fibrous investment of the liver.

The **Bile-secreting Apparatus** consists entirely of glandular cells (acini) and blood-vessels. The cells occupy the lobules, and are from about 1-1100th to 1-900th of an inch in diameter, each cell having a distinct nucleus. The cells are arranged in rows or columns within the lobule, and separated from the capillary vessels by a filmy membrane. At the periphery of the lobule this membrane coalesces and becomes continuous with the lining of the intralobular bile-ducts, which form a network at the sides of the cells. The liver is supplied with blood from two sources simultaneously—the hepatic artery and portal vein (Fig. 11).

The **Hepatic Artery** is a branch of the *cœliac axis*. It supplies nourishment to the liver, to the walls of its vessels (ducts, arteries, veins, and lymphatics), and the arterial blood from which, principally, the hepatic cells secrete the bile. The red blood-corpuscles, which were decomposed in the spleen, then pass to the portal vein, furnish, in their oxygen, hæmoglobin, and other elements, important constituents to the composition of the bilirubin and salts of the bile.

The **Portal Vein** is formed by the union of four large veins— the *gastric, splenic, superior mesenteric,* and *inferior mesenteric*. The portal vein conveys to the liver the venous blood from the stomach, spleen, pancreas, and large and small intestines. It also supplies the liver with glycogen, albuminose elements, organic and inorganic salts.

The glycogen is transformed into liver-sugar. In the substance of the liver the portal vein subdivides after the manner of an artery. Its branches run between the lobules, and are termed interlobular branches. They surround the lobule and anastomose. The portal vein has its origin and termination in capillaries. Its capillaries in the liver communicate with the intralobular or central veins, situated in and passing through the central axes of the lobules.

Fig. 11.

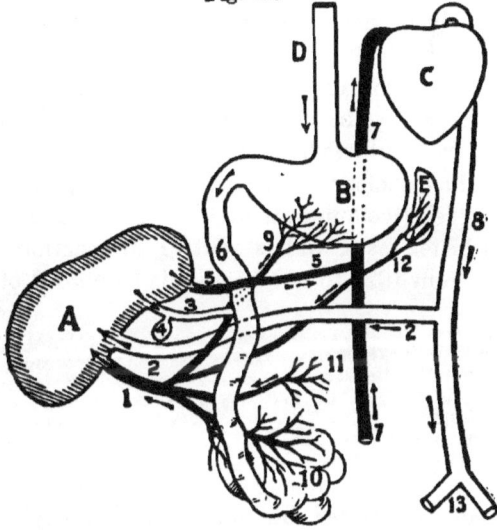

Diagram of the Hepatic Blood and Bile Vessels.

(The organs in the diagram are separated, in order to show the vessels. In the natural position, the liver partly overlaps the stomach; and the hepatic artery is a branch of the cœliac axis. This diagram is simply to indicate the four principal vessels entering and departing from the liver.)

A, Liver. B, Stomach. C, Heart. D, Œsophagus. E, Spleen.

1, Portal vein. 2, Hepatic artery. 3, Ductus communis choledochus. 4, Gall-bladder. 5, Hepatic vein. 6, Duodenum. 7, Inferior vena cava. 8, Descending aorta. 9, Capillary veins of the stomach discharging into the portal vein. 10 and 11, Capillary veins of the intestines, emptying into the portal vein. 12, Capillary veins of the spleen, discharging into the portal vein. 13, Iliac arteries.

From these central veins and from the arterial capillaries the rootlets of the hepatic veins arise. Thus an artificial injection into either the hepatic artery or portal vein is apparent in the hepatic veins. Some hepatic cells have the function of secreting bile, while others transform glycogen into sugar; thus the blood, as well as the nutritive materials, in passing through the liver, are changed. The albuminose elements absorbed by the capillary veins (mesenteric veins) of the intestines differ from the albumen in the hepatic veins. Furthermore, by the secretion of bile, the non-nitrogenous portion of the arterial blood is removed, just as the effete nitrogenous portion is separated and given off by the kidneys. The liver protects the body; substances just absorbed, but not fit for the circulation, are by the liver destroyed and excreted.

The liver has a receptacle (storehouse) called the **Gall Bladder,** in which the bile temporarily accumulates. During the process of digestion the bladder contracts and discharges the bile into the duodenum. That glycogenous materials undergo important changes in the liver, may be inferred from the fact that the hepatic veins contain more pure blood-sugar than there is of glycogen in the portal vein. The hepatic veins contain also more white blood corpuscles than the portal vein but less hæmoglobin, albumen, fibrin, and salts.

Pure Bile has a clear, yellowish-brown or greenish color, depending on the relative amounts of bilirubin and biliverdin present. It has a bitter taste, and is neutral in reaction. Its specific gravity ranges from 1020 to 1025. Bile is composed of:—

```
Water...................................................859.2
Solids—Bile salts (organic).............................91.5
        Inorganic salts.................................. 7.7
        Fat............................................. 9.2
        Cholesterin..................................... 2.6
        Mucus and coloring matters.....................29.8—140.8
                                                        ——————
                                                        1000.0
```

Bile is continuously secreted, but only completely when, at proper intervals, food enters the stomach and digestion goes on in the alimentary canal. During long-continued fasting some of its material is wanting, in which case the secreted bile cannot be entirely normal. The secretion is most abundant during digestion, when absorption of food in the intestinal canal has begun. The bile in the capillary bile-ducts and the blood in the capillary blood-vessels are conveyed by capillary peristaltic action. (*Vide* Capillary circulation in chapter on Circulation.) The flow of blood and bile in the larger hepatic vessels is assisted by the respiratory movements of the diaphragm and abdominal muscles, compressing and relieving alternately the portal and hepatic veins, thus favoring the functional activity of the liver. The two hepatic veins convey the blood and sugar from the substance of the liver, and carry it into the inferior vena cava; while the bile ducts collect and pass the bile into the hepatic duct, thence into the duodenum.

The daily secretion of bile in an adult is from about one and a half to two pints. It is undoubtedly antiseptic to the contents of the intestine, as when its contents are not acted upon by the bile the feces have a putrid odor. Bile is decomposed in the small intestine and gradually diminishes in quantity; its salts are absorbed, so that normally in the large intestine no biliary salts are found. Obstruction of the bile ducts leads to extreme costiveness, showing that bile is a natural purgative. An abnormally increased quantity of bile causes diarrhœa. When no bile enters the intestines the dissolution, emulsifying, and absorption of fat are incomplete, and the want of bile salts in the blood results in impaired nutrition and consequent reduction in the weight of the body.

The mucus secreted by the cells of the mucous membrane in the walls of the bile-ducts gives to bile a sticky consistency. Any slight resistance to the outflow of the bile from the liver, such as catarrh of the bile-ducts, or a superabundance of mucus on the mucous membrane of the intestinal canal, extending to the hepatic ducts, causes *biliousness*. If the obstruction be more serious, though the bile may be properly secreted, and its discharge merely interfered with—as by a plug of mucus, a gall-stone, or a tumor, which increases the pressure within the ducts, and enlarges the liver— then the lymphatics of the liver (not the veins) absorb the bile, which then passes by way of the lymphatics to the thoracic duct and into the blood, resulting in *jaundice*. This condition is always accompanied by extreme constipation, owing to the want of bile in the intestines. When bile has entered the blood, the pulse is generally below normal—60 to 40 beats per minute—accompanied by slow respiration, and the temperature is lower than normal. This condition is caused by the bile in the blood weakening the sensitiveness of the endothelial cells of the internal lining of the heart, so that stimulation by the carbonic acid of the blood on these cells is diminished, and the result is insufficient stimulation for nervous reflexes on the heart. The administration of veratrin or aconitin, especially the former, acts also on cells by diminishing their sensitiveness, just as an ether-spray on the skin diminishes the sensitive function of the *organum tactus*.

If the solid constituents of bile, especially the cholesterin, be in excess in the liver, and solidify, biliary *calculi*, or gall-stones, may be produced. At birth, by the ligation of the umbilical cord, the hæmoglobin in the ligated vessels is converted into bilirubin, generally producing jaundice in the infant for the first few days.

In health the glandular cells of the liver contain oil, in small globules. If in disease they accumulate to excess, the result is known as *fatty degeneration* of the liver. When bile enters the

5

stomach, it results in vomiting, as the bile is incompatible with pepsin. In administering hepatic stimulants, it is well that the bowels be kept open; otherwise, the bile, as such, may be reabsorbed in the intestine.

Hepatic remedies are generally absorbed by the capillary veins of the stomach. These veins discharge into the portal vein. Landois notes the influence of drugs on the secretion of bile in dogs: "Podophyllin, aloes, colchicum, euonymin, iridin, sanguinarine, ipecacuanha, colocynth, sodium phosphate, phytolaccin, sodium benzoate, sodium salicylate, dilute nitro-hydrochloric acid, ammonium phosphate, and mercuric chloride (corrosive sublimate) are all more or less powerful hepatic stimulants, through their direct action on the hepatic cells; while scammony, colocynth, jalap, sodium sulphate, and baptisin, act on both the liver and the intestinal follicles and glands. Among substances which stimulate these follicles and glands, but not the liver, are castor oil, calomel, magnesium sulphate, gamboge, ammonium chloride, manganese sulphate." Acetate of lead directly depresses the biliary secretions.

. The action of bile in the economy is five-fold, viz.—

1. To assist the process of emulsifying fatty matter in the intestines.

2. To assist in the absorption of fats by the lacteals of the villi.

3. To moisten and additionally stimulate the intestinal walls to peristaltic movements.

4. Its *taurocholin* and *glycocholin*, when absorbed in the intestine, furnish the blood with most important atomic elements. When they are not so absorbed the body emaciates.

5. Its antiseptic properties prevent putrefactive changes in the intestines. (*Vide* Ptomaines.)

Liver Sugar.—That part of starch which is not transformed into glucose by digestion is changed into glycogen, which becomes sugar as soon as it reaches the liver. That sugar is there formed is proven by the fact that if in the portal blood which supplies the liver no sugar appears, yet if supplied with glycogen sugar is found in the hepatic veins leading from the liver. Sugar is, therefore, formed by the transformation of starch during digestion, and from glycogen in the liver.

Sugar in Blood.—The proportion of sugar in normal blood is about two parts sugar in a thousand parts of blood, and possibly a portion of this, in passing through the liver unchanged, may be the sugar found in the hepatic veins when no glycogen enters the liver, as claimed by some physiologists. (*Vide* Sugar in chapter on Nutrition.) In the normal condition of the system a limited quantity of sugar undergoes changes in the blood, and is entirely decomposed; but if from any cause its quantity is abnormally increased, it is then found in the urine. A *diabetic* condition may result from an unusual amount of sugar being taken with food, or from anything which hastens the circulation of the blood through the liver, or from an oversupply of blood to the liver, as in continued struggles, convulsive muscular action, forcible compression of the abdominal organs, or injuries. Again, if the circulatory centre at the floor of the fourth ventricle in the medulla oblongata be abnormally irritated, it increases the nervous reflex impulses to the abdominal and hepatic circulation, and sugar may then be found in quantity in the blood and urine. This condition may, on account of some disturbance in the circulation, be only temporary; if permanent, it is called *diabetes mellitus.* This latter affection is generally progressive and fatal, the urine being increased in quantity, of abnormally high specific gravity, and continuously charged with sugar, which may continue to appear even when no saccharine matter is taken with the food.

The **Nerve-fibres supplying the Liver** are derived from the *hepatic plexus*, which is formed by nerve-fibres from the cœliac plexus of the sympathetic nervous system and some fibres from the left vagus and right phrenic nerves. The fibres of the hepatic plexus accompany the hepatic blood-vessels, and supply their walls and the secretory cells. The normal sensitive stimuli for continual bile-secretion is derived from the carbonic acid of the venous blood in the liver. This sensation is conveyed through the vagus fibres to the medulla oblongata, where it is reflected and efferent (motor) impulses are conveyed through the phrenic nerve back to the liver for its action—similar to the reflex stimuli to the lungs or heart. Artificial stimuli or depressors (medicines affecting the liver) may increase or diminish these hepatic actions. (*Vide* Carbonic Acid in chapters on Respiration, and Circulation.)

The **Lymphatics of the Liver** form both superficial and deep networks. The superficial lymphatics are found in the tissue beneath the mucous membrane of the bile-ducts; the deep accompany the blood-vessels to the sides of the hepatic cells.

Pancreatic Juice.—The **Pancreas** is a lobulated gland (A, Fig. 12). In structure it is like the salivary glands, but much larger. It lies transversely in the upper part of the abdomen; its right end is in contact with the duodenum (C). Its ducts (1) run through it centrally from end to end, receiving by their branches the secreted pancreatic juice and emptying it (2) into the duode-

num, just below the pylorus. The biliary duct (3) from the liver
also enters the duodenum, a little above the orifice of the larger
pancreatic duct. In this manner the bile and the pancreatic juice
are mixed with the chyme soon after it enters the duodenum.

About ten ounces of pancreatic juice are secreted daily by an adult

Fig. 12.

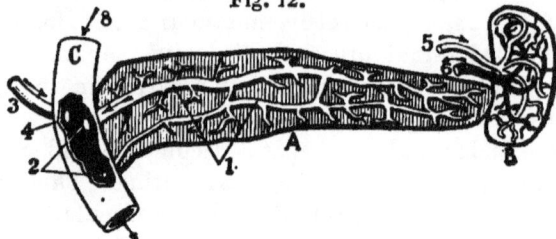

Diagram of the Pancreas and Spleen.

A, Pancreas. B, Spleen. C, Duodenum.
1, Pancreatic ducts. 2, Orifices of the pancreatic ducts. 3, Bile
duct. 4, Orifice of the bile duct. 5, Splenic artery. 6, Splenic vein.
7, Capillary vessels in the spleen. 8, Indicating the passage of food
from the stomach to the duodenum.

during the digestive processes; but none in the intervals between.
Its principal chemical constituent is an albuminoid substance
termed *pancreatin*, constituting about ten per cent of the pancreatic
juice. Zamanski, who made a careful examination of healthy
human pancreatic juice, reported it to be a somewhat tenacious,
yellowish, turbid liquid, with a marked alkaline reaction.
His analysis gave the following results:—

"Water... 86.50
Organic compounds (including albuminoid substances 9 2-4ths,
    extractive matter 3 3-4ths. Of the latter nitrogen constitutes
    about 1 1-12th part)... 13.25
Salts (sodium carbonate, chloride, phosphate, and sulphate;
    potassium, calcium, and iron), about...................... .25
                                                              _____
                                                              100.00

At a temperature of 100° F. (38° C.), this pancreatic juice readily
converted starch into glucose, egg-albumen into peptone [albu-
minose ?], and olive oil into an emulsion." (*Vide* Peptone in
chapter on Digestion.)

Fatty materials are not changed in the stomach; but melted by
its warmth, and set free from the vesicles and fibres of the food.
As soon as this fat comes in contact with the pancreatic juice, the
emulsifying process takes place, after which it is ready to be absorbed by the lacteals and to pass into the circulation. The pancreatic juice also converts into glucose those starchy substances
that escaped the action of the saliva and passed unchanged into

the duodenum, the pancreatic juice possessing this power in a far higher degree than saliva, thus preparing starchy matter for the final sugar refinery—the liver. (*Vide* Starch and Fats in chapter on Nutrition.)

Vessels of the Pancreas.—The pancreas is supplied with arterial blood from numerous branches of the splenic, pancreatico-duodenal branch of the hepatic, and superior mesenteric arteries. The veins discharge into the splenic and superior mesenteric veins, which are tributary to the portal vein. The lymphatics communicate, eventually, with the thoracic duct.

Nerves of the Pancreas.—The nerve-fibres of the pancreas are from the solar and splenic plexuses. Its afferent fibres run within the pneumogastric nerve, its efferent in the phrenic nerve.

The **Spleen** (E, Fig. 11; and B, Fig. 12) is a ductless gland, concave and oblong in form, soft, and of a dark bluish-red color. Considering its size, it is the most vascular organ of the body. In the adult it is about five inches in length, four in breadth, one and a half in thickness, and weighs about seven ounces. It is situated in the left hypochondriac region; its upper portion is in close proximity with the diaphragm. It is held in position by a reflection of its peritoneal covering, termed the *suspensory ligament.* Its upper inward surface is in close contact with the cardiac portion of the stomach, adhering to it by the *gastro-splenic omentum.* Its lower inward surface is in close contact with the pancreas. The spleen is invested with two coats, a serous (peritoneal) externally, and a fibrous elastic coat beneath. Its nutritive supply is derived from the blood of the splenic artery. Its venous blood is conveyed by the splenic vein to the portal vein, which conveys it to the liver.

The pulp of the spleen is made up of blood-corpuscles, granular matter, nucleated bodies (splenic corpuscles), and colorless nucleated cells in size from 1-7000th to 1-6000th of an inch. These, together with the intercellular membranes, form its parenchyma. Numerous lymphatic vessels arise from the proximity of the splenic corpuscles.

Physiologically, the spleen is not absolutely necessary to life. It has often been removed from animals, and by accident destroyed in man, without perceptible loss of health after the wound had healed. Still, its functions are manifold. It has been surmised that the spleen is: 1. A *diverticulum*, or reservoir, of nutritious elements for the blood, to be used during fasting. 2. The broken-up red blood-corpuscles impart the separated elementary substances to the venous blood of the portal vein. 3. To be a storehouse for albuminous matter, to be drawn upon and used during fasting. 4. To have the function of selecting and aggregating fibrin elements, as shown by the blood of the splenic vein, which sometimes contains four to six times the quantity of fibrin found in the venous blood elsewhere.

That the spleen produces new blood-corpuscles is a conjecture not yet confirmed. The fact that the white corpuscles found in the vessels leading from the spleen are in excess of those found in the vessels supplying the spleen, may be accounted for by assuming that one function of the spleen is the changing of some of the red corpuscles into white corpuscles. The hæmoglobin and oxygen released from the red blood-corpuscles in the spleen pass on to the liver, and there enter into the formation of bilirubin, that constituent of bile which gives it its characteristic color.

The parenchymatous intercellular structure of the spleen allows great extension, forming *diverticula*, thereby serving to relieve the portal vein from undue distension under various circumstances. When the secreting function of the liver is enfeebled, or its ducts more or less blocked up, the spleen is always found enlarged. If the intestinal canal be distended with food, it may cause pressure on the portal vessels; then the speen enlarges. Again, it is large and distended after the process of chymification—say, from one to three hours after food is taken. It is manifest, therefore, that the spleen furnishes the liver with elements necessary for the production of bile, not only during the process of digestion, but also during the periods when no food enters the body. In prolonged fasting the spleen becomes correspondingly small, showing that it gives off its superfluous accumulation of blood to the liver to furnish material necessary for the continuous secretion of bile, as necessity requires. The spleen is, therefore, a reservoir for the blood, the quantity of which is increased when food is taken; for the superabundance of nutritious elements to the blood entering the blood-vessels at once, would endanger health if no provision were made for a storehouse—the spleen. As soon as the absorbing and secreting organs have relieved the general blood pressure of the system, the spleen gives up its surplus, and regains its previous dimensions. This indicates that, if the spleen be removed, or its power of distension be impaired, there is liability to *apoplexy*. Such a condition may be relieved most quickly by venesection. When from any cause the peripheral circulation becomes enfeebled (chills), or there is a functional derangement of the liver, as in malaria, the spleen enlarges.

CHAPTER VI.

SECRETION AND EXCRETION.

Secretion is the function, possessed by certain classes of cells, of absorbing atomic elements from the blood, and transferring and combining these elements into one or another of four forms of new compounds — solids, semisolids, fluids, or gases. The *solids* we find as collagen, chondrin, etc.; the *semisolids* as mucosin, myosin, etc.; the *fluids* as saliva, gastric, pancreatic, and intestinal juices, bile, tears, perspiration, urine, milk, and serous fluids, etc.; the *gaseous* as carbonic acid gas in expired air. Secretion is effected solely by cells, as certain cells of bones, muscle-fibres, membranes, follicles, glands, and walls of ducts. The products of some of these secreting cells are deposited in the tissues, as the collagen, chondrin, and myosin; others are for immediate action, as the saliva, gastric and intestinal juices and fluids; others, again, are excreted, as urine and perspiration. Some of the secreting-cells are somewhat isolated, as those of bones and of muscle-fibres; others side by side, as those of mucous and serous membranes, Lieberkühn's follicles, and ducts; others, again, in grape-bunch form, as those of Brunner's glands, the pancreas, and salivary glands. The secreting glands are divided into two great classes; viz., those with ducts, and those without. To the former glands belong the lachrymal, salivary, sudoriparous, mammary, sebaceous, Brunner's, the liver, pancreas, kidneys, and testes; while the latter include the suprarenal capsules, spleen, thymus, thyroid, pituitary, and solitary glands. Some secreting glands are racemose in form, or grapebunch-like, such as the salivary, and Brunner's; others appear as a solid organ, as the liver and kidneys. In most cases the secreting cells derive the materials necessary for their own nourishment and for the elaboration of their ultimate products from the arterial blood only; while others derive the materials necessary for their secretions from the venous blood also, as the secretions of the cells of the liver, lungs, kidneys, and skin. The various secretions are not found, as such, in the blood, but are products of the activity of the protoplasm of the cells, attracting, selecting, and aggregating

the necessary atoms and molecules from the blood, and thus forming, according to the peculiarity of each class of cells, the various
compounds—secretions. Hence, the phenomenon of secretion and
the formation of new products is a function of the cells, and not a
mere filtration from the blood.

The secreting organs are supplied with two classes of nerve-fibres.
These fibres are alike in structure, but carry different kinds of impulses; viz., sensory and motor. The *sensory* nerve-fibres convey

Fig. 13.

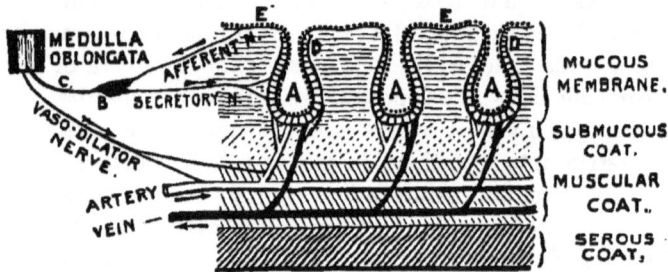

Diagram of Secreting Follicles.

A, A, Follicles. B, Ganglion of the sympathetic system. C, Nerve-
fibre communicating with the medulla oblongata. D, D, walls of the
follicles containing columnar epithelial cells secreting the principal
active substance of digestive juice. E, E, Epithelial cells lining the
mucous membrane.

impulses inward (afferent) from the secreting organ to the medulla
oblongata, which, by reflex action, sends out an impulse (efferent)
through one or more motor fibres to the appropriate organ or
organs. The *motor* nerve-fibres carry four kinds of impulses; viz.,
vaso-motor (constrictor and dilator), inhibitory, secretory, and
trophic. The impulse through the vaso-motor regulates the distribution of blood in the capillaries of the organ; the impulse through
the inhibitory fibres diminishes or suppresses its activity; others
regulate the activity of the secretory cells; and still others, the
trophic, regulate the nutritive activity of the organ. Transmission
of impulses through the three first mentioned motor-fibres takes
place during secretory activity only, while the fourth, or last, is
most active during the repose of the organ. That the transmission
of impulses through each class of nerve-fibres named is independent of that of every other class, is proven by the fact that one kind
of impulse may be defective or incomplete, or the nerve-fibre paralyzed, while the others remain intact and unimpaired. Again, a

secreting organ may become congested, and lose the secreting power, though, normally, secretion is carried on when the organ is more or less congested. The motor fibres (vaso-motor, inhibitory, secretory, and trophic) from the medulla oblongata pass directly through, branch from, or reticulate with the sympathetic plexuses, and are thereby enabled to communicate simultaneously with one or many secreting organs.

The impression received from a stimulus on the end-organ of a sensory nerve in a secreting organ is produced by contact with foreign matter, though the individual may not be cognizant of it, as the cerebral convolutions are not directly connected with the function of any secreting organ. The medulla oblongata is thus connected, however, so that whenever a secretory organ becomes active or inhibited by *mental* influence, the mind can act only reflexly through the medulla, similar to stimuli from a secreting organ. Consequently, any form of stimulation from the mind, for secretion or its impairment, must be considered a reflex act through the medulla. The presence of food or a foreign substance in the mouth, or even the smell of a favorite dish, provokes the flow of saliva; the presence of this food or substance in the œsophagus, stomach, and intestines, stimulates their secretions, as well as, indirectly, those of the liver and pancreas; while the presence of a very hot drink in the œsophagus and stomach, or mental excitement or depression, may cause perspiration. The afferent nerve-fibres from secreting organs pass through sympathetic ganglia before arriving at the medulla oblongata, similar to the manner in which all afferent (sensory) fibres pass to cerebro-spinal centres; while the efferent (motor) fibres from the medulla oblongata to secreting organs all pass to or through sympathetic plexuses before entering the organs. The activities of the elements, cells, and tissues cause a rise of temperature in the secreting organ, venous blood issuing from the organ, and in the newly formed product. The temperature, therefore, is higher in them than the temperature of the arterial blood supplying the organ.

There are four fluids in the body with an acid reaction; viz., gastric juice, urine, perspiration, and the mucus of the vagina; all others have an alkaline reation; *i. e.*, the blood-plasma, lymph, milk, tears, saliva, mucus, the serous, spermatic, intestinal, and pancreatic secretions. A characteristic of all secretions that are *useful in the economy* is that they have an alkaline reaction, except the gastric juice, and even this becomes neutralized; while those which are acid, and so remain, must be removed, as their accumulation in the body would prove deleterious to health.

Excretion (*evacuatio*) is the separation, emission, or expulsion from the body of materials which have become useless, constituting the effete matters or waste products (except the feces) of the various chemical changes and decomposition of the tissues. These materials, which have been a part of the tissues, by continual changes are no longer useful to the system, hence are collected by the absorbent vessels (lymphatics and veins), and carried by the blood to proper places for elimination; for example, the urine, carbonic acid gas, and perspiration. The salivary and mammary glands are secretory in their function, while the lungs, liver, kidneys and skin are secretory as well as excretory organs. The urine is secreted by the cells of the tubuli uriniferi in the cortical portion of the kidneys, but its expulsion takes place through the urethra. It is claimed by some that the separation of urine from the blood is simply a filtering process brought about by the blood's high pressure in the glomerules of the kidneys. This cannot be altogether true, however, for the following reasons, given by Yeo, to-wit:

"1. Urine differs much from the fluid obtained by filtering the blood.

"2. In health, the urine contains no albumen, a substance in which the blood is rich.

"3. Urine contains comparatively much more urea and salts than the blood.

"4. Why should simply water and salts, without albumen, pass through the capillaries of the glomerules and not through any other capillaries?" (For further demonstration see chapter on Kidneys.)

The due performance of the functions of excretory organs is essential to the well-being of the body, for the following reasons: The peculiar anatomical and chemical composition of the tissues of the body renders them liable to continuous decomposition and decay. Consequently, it is a positive necessity that the decomposed matters be rapidly carried off. Every physical, mental, and chemical operation causes disintegration, or breaking up, of certain portions of the tissues, resulting in the accumulation of effete matter, which, with the decomposed food and food excess, would, if not speedily removed by the excretory organs, cause the blood soon to become poisoned.

CHAPTER VII.

THE ABSORBENT SYSTEM.

The absorbent system consists of capillary veins and lymphatics. The absorbents take up materials from the intestines and tissues, and carry them into the general circulation. (The **Venous System** is described in the chapter on Circulation.)

The **Lymphatic System** consists of vessels, glands, and the thoracic duct. The vessels ramify and extend throughout nearly all parts of the body, and possess the function of absorbing the chyle from the intestines, and the lymph from other organs and tissues, conveying them into the venous system. Lymphatics lead, like veins, towards the heart. The lacteal vessels, originating in the villi of the intestines, pertain to and are a part of the lymphatic system. They absorb and convey the chyle to the follicles beneath the villi, thence pass through lymphatic glands, and discharge into a much-dilated lymphatic vessel known as the *receptaculum chyli* (1, Fig. 14). This dilatation is the commencement of the thoracic duct, and is formed by the junction of the two lumbar and intestinal lymphatic trunks, just in front of the second lumbar vertebra, between the abdominal aorta and inferior vena cava. The lacteals are so termed because the chyle, which they absorb from the alimentary canal, is milky in appearance.

Chyle is the product of digestion. It is found in the small intestine and the lacteals. In order that digested food may be of use to the entire economy the nutritious materials pass through the mucous membrane of the alimentary canal, so as to obtain admission into the blood. Its endosmosis through the villi and mucous membrane is facilitated by the alkaline condition of the intestinal juices. It passes into the thoracic duct and mingles with the white lymph, entering the general circulation at the left subclavian vein. The mucous membrane of the intestine, and the vessels therein, are lined with cells endowed with the function of controlling the passage of liquids and liquid foods through them, and of elaborating new juices. Consequently, these cells are divided into two classes— one for secretion, the other for absorption. The former are the cells of the follicles of Lieberkühn and glands of Brunner. These absorb from the arterial blood the material for the secretion of the intestinal juices, which are discharged into the intestinal canal. The cells of absorption are those of the intestinal capillary veins and lacteals, which absorb and promote the passage of nutritious substances from the intestines to larger veins and lacteals.

Fig. 14.

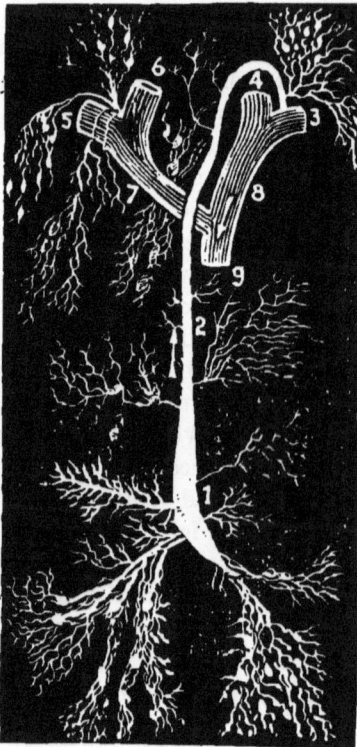

Diagram of the thoracic duct.

1, Receptaculum chyli. 2, Thoracic duct. 3, Left subclavian vein. 4, Left internal jugular. 5, Right subclavian vein. 6, Right internal jugular. 7, Right vena innominata. 8, Left vena innominata. 9, Superior vena cava.

The oleaginous matter of the chyle is in a state of emulsion; i. e., suspended in minute particles in the nutritive fluid, like fine granulated dust. These fine fatty granules constitute the *molecular basis* of the chyle. The white color and opacity of chyle depend upon the molecular condition of the oily ingredients. During digestion the capillaries of the lacteal vessels become distended. As soon as the digestive process is over, and the contents of the lacteals are discharged into the thoracic duct, the lacteals again contract. Chyle is not therefore constantly present. The products of digestion are carried by two different routes—by the portal vein, and by the lacteals—into the general circulation. The vena porta carries the blood from the portal system, which contains glycogen, to the liver, where it traverses the capillary vessels and is transformed by the cells of that organ, and the product is conveyed by the hepatic veins to the inferior vena cava, thence to the right auricle of the heart. The lacteals absorb the chyle and convey it to the thoracic duct, which carries, besides the chyle, the lymph of the trunk and lower extremities. The chyle contains the largest proportion of albuminose, fibrinose, and fatty ingredients derived from the ingesta. The thoracic duct contains, besides the lymph and chyle, the white blood-corpuscles derived from lymphatic glands. The latter are especially numerous on the lacteal vessels of the mesentery, through which the chyle flows on its way to the thoracic duct. All these substances become modified after entering the blood, and, except the water and some of the salts, are no longer recognizable under their original form.

Water may be absorbed from the intestines in an almost unlimited quantity. A weak solution of common salt may become partly absorbed; but if salt elements be present in the solution in great quantity, as in large doses of sulphate of magnesium, absorption is then stopped, and even reversed. In this case, instead of the fluid parts of chyle passing inward by endosmosis, exosmosis takes place, causing thereby copious watery stools. With fat there is a similar phenomenon. A small quantity is absorbed by virtue of the alkalinity of the intestinal juices and the presence of pancreatin; if however, fat is present in excess, it acts as a laxative and passes off with the feces.

Lymph is the fluid in the lymphatic vessels absorbed from the tissues of the body. It is a whitish yellow, transparent, juice-like liquid, salty to the taste, alkaline in reaction, and having a specific gravity of 1022. It is composed of water, fibrin, albumen, fat, and salts. It coagulates like blood when withdrawn from the vessels; that is to say, on account of the fibrin present, it separates into two parts—fibrin (a solid), and serum (a liquid). The same propelling influences are brought to bear on the lymph as on the venous blood. In the average adult, about sixteen and a half ounces, or about 2000 grammes, are daily absorbed and conveyed to the blood. Lymph is derived from the metamorphosis of the tissues, and passes into the circulation, to be transformed in part into other matter, the residue being eliminated from the system by excretion. The lymphatic vessels are very thin, delicate, and transparent, nearly uniform in size, and somewhat nodulous, on account of their numerous valves—being similar in valvular structure to veins. These vessels are arranged in two sets, the superficial and the deep. The former are found immediately beneath the skin, and join the deep set in the submucous and subserous areolar tissues. The deep lymphatic vessels accompany the larger blood-vessels. The lymphatic capillaries originate in the tissues by plexuses, and lie between the plexuses of capillary blood-vessels. There the various fluids transude from one set of vessels to another set; the arterial blood nourishing the intervening tissues, while at the same time the veins take up the blood, and the lymphatic vessels absorb the worn-out material. The molecular activities of the tissues, and the alternate contraction and relaxation of the muscular fibres during muscular activity increase the metamorphosis of the tissues and the quantity of lymph.

Lymph Glands.—The lymphatic capillary vessels enter lymphatic glands, which vary in size from that of a small pea to that of a small walnut, and are situated in the course of the lymphatic vessels (Fig. 14). Lymph contains white blood-corpuscles, formed in the lymph glands, and are carried away by the stream of lymph flowing through the glands. These glands are very numerous in the neck, axilla, chest, abdomen, and especially in the mesentery and groin. In the latter situation the glands sometimes become inflamed, and *bubo* results. Lymphatic vessels are found in nearly all parts of the body, the exceptions being the cornea, the vitreous humor of the eye, and the peripheral epithelial coverings.

The chyle of the abdomen and the lymph of the left side of the body and arm, and of both legs, are conveyed to the thoracic duct, thence into the left subclavian vein (3, Fig. 14); while the lymph of the head, neck, and right side of the chest and arm, enters the circulation at the junction of the right subclavian and right internal jugular veins (5, Fig. 14). In this manner all the absorbed substances of the body are collected, and become mingled with the venous blood, just before entering the right side of the heart. (*Vide* Intestinal Digestion.)

Osmosis is the interchange of liquids through organic membranes or tissues. **Endosmosis** is their passage from without inward; while **Exosmosis** is their passage from within outward. In the living body, the membranes and tissues are all fresh; and with the presence of fluid-matter containing sodium chloride, and the proper normal temperature, the conditions for endosmosis and exosmosis are present. Egg-albumen is non-osmotic; *i. e.*, it does not pass by transudation through an animal membrane, on account of its adhesiveness and gelatinous consistency; nor does the albuminous matter of the blood exude into the secreted fluids without being acted upon catalytically and dialytically. A certain pressure is necessary to obtain the result. If the dialytic pressure in the capillary arteries be increased, while obstruction to the venous capillaries exists, the albumen transudes with the saline and watery parts of the blood, infiltrating the tissues of the affected part. In this way a collection of blood-serum in the cellular tissue under the skin takes place, called œdema, a general *anasarca*, or *dropsy*.

CHAPTER VIII.

THE BLOOD.

The **Blood** is a red liquid that circulates through the cavities of the heart, arteries, veins, and capillary vessels. It consists of the *liquor sanguinis*, with minute red and white bodies, or *corpuscles*, floating independently in the liquor sanguinis in large numbers. The corpuscles form by weight about 45 per cent and the plasma about 55 per cent of the blood, or by volume about 60 and 40 per cent, respectively. The specific gravity of the blood is 1055. It is slightly alkaline.

Red Corpuscles.—Arterial blood contains about 6 millions of red corpuscles in each cubic millimetre. In form the corpuscle is a spheroidal biconcave disc, in size from about 6 to 8 micromillimetres (1-3200th of an inch) in transverse diameter (1 and 2, Fig. 15), and about 1-6th of the transverse diameter in thickness (3). It has no external envelope and no nucleus, and is composed of water, albuminous and coloring matter, iron and mineral salts. It is jelly-like in consistency, and very flexible, and therefore easily elongated in the course of circulation, capable of twisting, bending, and passing through narrow and tortuous channels. In fact, it will pass through vessels having a smaller diameter than its own, regaining its original shape as soon as the pressure or traction is removed. Its form is determined principally by external influences; *i. e.*, by imbibition of oxygen, fluid, or by tissue pressure. If water be added to the blood the red corpuscles absorb it by imbibition, whereupon they lose their concavity, and become pale and spherical. Strong alkaline solutions, such as sodium, potassium, or ammonium, dissolve the whole substance of the corpuscle with more or less facility. After circulating for a time through the capillaries they become dissolved in the liquor sanguinis. The red corpuscles, when examined shortly after being drawn from the vessels, adhere together by their surfaces, and appear like rolls of coin or rouleaux (1, Fig. 15). The white corpuscles, by an inherent power, constantly undergo change of form (4), in the same manner as the amœbæ.

The **Hæmoglobin,** or red coloring matter, is the principal constituent of red blood-corpuscles, making up about 90 per cent of their substance. It is an albuminoid (proteid), completly formed within the corpuscles, possibly during the circulation of the blood through the right ventricle of the heart and through the capillaries of the lungs. The hæmoglobin of arterial blood consists of: Carbon, about 55 parts; oxygen, 21.85; nitrogen, 14; hydrogen, 8; iron, 1; and sulphur, 0.15; including the two compounds *globulin* and *hæmatin*. During its circulation through the capillaries of the tissues its chemical constituents change, so that the hæmoglobin in the corpuscles of veins near the capillaries consists of carbon, about 70 parts; oxygen, 4; nitrogen, 18; hydrogen, 8. These changes of constituents and color are effected during the blood's cir-

Fig. 15.

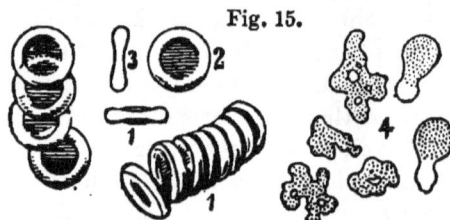

Diagram of Human Blood-corpuscles.

1, Red corpuscles lying in rouleaux. 2, As they appear when lying on their flat surface. 3, As seen from the side, exhibiting their double-concave form. 4, White corpuscles (leucocytes) exhibiting their changes of form and movements, as the amœbæ.

culation through the capillaries of the lungs and the tissues. Hæmoglobin is a powerful absorbent of oxygen when in contact with the atmosphere. The bright red color of the corpuscles depends upon the quantity of hæmoglobin and oxygen present. When pure hæmoglobin is separated from the corpuscles and dissolved, it may be crystallized. It is dissolved hæmoglobin which imparts to muscular tissue its reddish color.

The principal function of the red corpuscles is to carry oxygen, which they hold in loose combination. In certain of the lower animals these corpuscles are larger though less numerous than in the human economy; hence, their vital functions are lower. In animals whose red corpusles are small and numerous, the functions of respiration, circulation, nutrition, and motion are correspondingly increased. As the red corpuscles have no nuclei, they cannot be prepared by the blood; they are replenished by the transformation in the lungs of the young white corpuscles, which originate in the chyle, lymphatic glands, spleen (?), and red marrow of bones, continually passing into the great venous trunks. In the formation of red corpuscles a fresh supply of oxygen is absolutely essential. The constituents of the blood are present only temporarily, as they undergo chemical changes continuously in the different organs and tissues. In 1000 parts of healthy normal blood about 790 parts are water and 210 solids, the latter including the corpuscles (134 parts), albumen (52), salts (8), fats (5), sugar (2), fibrin (3), and waste matter (6). By evaporating the 79 per cent of water from the blood, there remain 210 parts, or 21 per cent of solid residue, the ultimate chemical composition of which resembles that of muscles.

When the red corpuscles are in excess in the blood, a pathological condition, or plethora, takes place, exhibiting itself in hæmorrhage or local congestion. Should they become deficient in number, anæmia results. To these corpuscles belong the special function of taking up oxygen from the atmosphere or air in the lungs. They contain an intercellular fluid substance of denser consistency than the serum, which, from the fact of these corpuscles being in the blood, is classified as tissue. This tissue, above all others, changes its chemical constituents and replaces its losses most quickly.

White Corpuscles.—Under normal conditions, the white corpuscles are not so numerous as the red, being about 1 to 300. They are about 1-4th larger than the red—*i. e.*, about 1-2500th of an inch in diameter—and are of irregular shape (4, Fig. 15). They consist of a soft, viscid, colorless, finely granular substance. Each white corpuscle has a nucleus, differing in this respect from the red corpuscle. A strong alkaline solution dissolves them. These corpuscles are endowed with *amœboid motion; i. e.*, they move very slowly, and sometimes stop for a few moments and adhere to the internal surface of the vessel. When the circulation of the blood is normal the amœboid movements cannot be seen; but if the blood be withdrawn from the vessel and kept at a normal temperature, they may be seen to change their shape, like the amœba, by alternate contractions and relaxations of their substance. Whenever the blood current in the vessels or tissues becomes obstructed or retarded, the white corpuscles accumulate, and become more numerous in proportion to the red ones in the part affected. It is believed that the white corpuscles originating in the spleen (?), red marrow of bones, and in chyle become red by the absorption of oxygen in the lungs, but that those originating in the lymph-glands remain white; also, that the white corpuscles can pass through the walls of capillary blood-vessels into the surrounding tissue, where they form corpuscles of pus (pus cells), or inflammatory lymph, and so play an important part in pathological processes.

6

A poison working in the body cannot always be found in the blood—*e. g.*, the poison of small-pox or vaccine lymph—though the pustules and lymphatics exhibit parasitic life. The presence of cocci in the blood in cases of measles seems to be an established fact, though the effect of an injection of measle-blood into a healthy person is yet in doubt; some experimenters have failed to produce a true measle disease, while others claim success. In this disease the number of red blood-corpuscles is diminished, while the number of the white is increased. The blood appears thin, dark, with diminished fibrin, and in fatal cases may infiltrate and stagnate in the tissues.

Blood-plasma is a transparent, colorless, homogeneous liquid, with blood-corpuscles held in suspension. In 1000 parts of plasma there are about 900 parts of water, 77 of albumen, 3 of fibrin elements, 8.5 of other nitrogenous and organic substances, 2.5 of fat, and about 9 of salts, such as sodium and potassium chlorides, carbonates, and phosphates, lime and magnesium phosphates. These salts are of importance to the system. They render the blood alkaline, a prerequisite to nutrition, as its alkaline condition enables the blood-plasma to neutralize the acid (a form of lactic acid) of the various tissues. This reaction results in the generation of carbonic acid elements, which are carried to the lungs, skin, and kidneys. In these organs the carbonic acid *gas* is secreted and eliminated. (*Vide* Carbonic Acid in chapter on Respiration.) The alkalinity of the blood also facilitates endosmosis, as well as the absorption of fatty substances, which could not otherwise take place, especially in the alimentary canal. For the purpose of nutrition the most important ingredients of the blood-plasma are the albuminous matters.

The serum of the blood is derived from the products of digestion and absorption. The blood-corpuscles, as well as the secreted juices are elaborated by fixed organs, and depend on them for the main-, tenance of their normal condition; it is important, therefore, that these fixed organs perform their functions perfectly, a disturbance in any organ changing the normal composition of the blood and leading to a perversion of function called *dyscrasia*. A dyscrasia of the blood itself is not common, the trouble arising generally in one or more of the fixed organs.

The blood's circulation diffuses the nutritious substances throughout the tissues, and conveys all the material to the parts where heat is generated, and supplies heat where lost by exposure. It also takes up and conveys waste products to the places of exit from the body.

The rate of movement of the blood current varies. In the heart, during its systole, or the contraction of the ventricles, its velocity reaches 1200 millimetres per minute; in the descending aorta, 300 to 600; carotid arteries, 200 to 400; venæ cavæ, 200; radial arteries, 100; and capillaries, 5. The average velocity of the blood in the arteries is from 20 to 30 centimetres per second. The rapidity of the circulation is such that the blood makes the entire round of the body in an adult in about 30 seconds. The amount of blood in the system is much greater during the time of digestion than during the period of abstinence. Normally, it is about 12 per cent, or 1-8th the weight of the body. In case of death from hæmorrhage about one-fourth of the blood remains in the body.

Each tissue absorbs from the arterial blood the particular materials it requires for nutrition, such as fibrin by the muscles, fatty substances by the nerves, gelatin and earthy salts by the bones, while the milk cells (during lactation) separate albuminous, fatty, and saccharine substances. Owing to the abstraction therefrom of the elements of nutrition in the tissues and secreting processes, the blood in each organ undergoes an alteration in its constitution, rendering it for the same purpose, unfit for further use. The venous blood concentration and intermixture in the heart is, therefore, rendered necessary to avoid deterioration and to maintain its normal composition.

Matters entering the blood may be quickly and completely eliminated, or more or less retained, even for years exhibiting it effects. This is illustrated in the introduction of the poisons of vaccine virus, syphilis, and of other chronic diseases. That this persistency is due to the blood carrying the morbid matter is proven by the fact that the same kind and parts of tissues become affected in organs on both sides of the body or limbs, the same kind of tissue in other parts very often remaining unaffected; also, by the existence of a certain adaptation, peculiar to each individual person, between the blood and the tissues. It seldom occurs that even the same kind of tissues become in all persons simultaneously, similarly, or as severely affected. Impure or bad blood depresses and weakens the nervous system especially, its gray matter being the most delicate of all tissues, and soonest affected by faulty or insufficient nutrition.

In the disease termed chlorosis the red blood-corpuscles are considerably reduced in number, the blood appearing watery, though really having no more water than usual, and being lighter in color than when normal. This chlorotic condition does not depend on the disease or decay of these corpuscles, but upon their imperfect and insufficient reproduction. In most cases chlorosis is a congenital disease, depending on dyscrasia. It occurs most frequently in young females, exhibiting itself in fatigue, palpitation of the heart, pains in the back and hips, fermentation and flatulence in the stomach and bowels, and constipation and an unnatural appetite for chalk, lime, and other absorbents—all the symptoms indicating a weakness of the digestive organs and general debility. As the disease advances the face and skin of the body appear greenish white, the feet become œdematous, and there is a condition of general anæmia.

Another disease simulating chlorosis is called leucocythæmia, which may more properly be termed *leucocythamplio*, indicating an abnormal increase in the number of white blood-corpuscles. The disease is commonly known as the *white-blood disease*. In chlorosis the blood appears watery, in leucocythamplio white. In the latter disease the relative proportions of red and white corpuscles is changed, so that instead of 300 there may be only 50, 20, or 10 of the red to one of the white, or they may even be equal in number. In this affection an increased production of white corpuscles in the spleen (?) and lymphatic glands takes place. This is shown by the fact that the veins and absorbent vessels which carry the blood and products from these organs contain not only more white corpuscles than under normal circumstances, but more also than the arteries that supply them. In leucocythamplio the spleen is always considerably enlarged (splenic tumor—hyperæmia). The blood in the vessels of the pulp of the spleen, as well as the contents of its lymphatic vessels, flowing more slowly, an accumulation takes place, and consequent swelling results. When the spleen has been distended for a time, an increase of new capillaries (hyperplasia), or enlargement of its entire structure, and overproduction of its white blood-corpuscles, follow. This leucocythamplic process, or swelling of the spleen, generally causes simultaneous enlargement of the lymphatic glands, which may lead to chronic inflammation of the peritoneal covering. The blood in malarial disease is deficient in albumen, fibrin, and red corpuscles, which explains the co-existing anæmic condition, palpitation of the heart, and œdematous condition of the skin. In this disease the liver and spleen are always enlarged, the enlargement of the spleen being brought about by the damming up of the blood in the portal vein and its tributaries. Again, the blood is defectively developed, owing to the deranged state of the stomach, bowels, and blood-making organs.

The coagulation of blood depends on the presence of fibrin. Fibrin, though present in such very small quantity (only 3 parts in 1000 of blood), exhibits its great importance. When blood is gradually withdrawn from the blood-vessel, it is this fibrin which causes its coagulation. Fibrin, however, does not exist in blood as such. There are two substances—the *globulin* of the corpuscles and *fibrinogen* in the plasma—which unite to form the fibrin of the coagulum. A portion of the fibrinous elements of blood disappears in passing through the capillary circulation, showing that venous blood contains less of these elements than arterial. In the liver and kidneys it entirely disappears, so that none is found in the renal and hepatic veins. This demonstrates clearly that it is produced elsewhere, and is always of recent formation, its quantity remaining unchanged. When a congenital deficiency of fibrinous elements exists in the blood, no clot can form in a ligated vessel; so that if a ligature be applied and afterwards removed, bleeding begins again. Another ligature is likewise followed by a secondary hæmorrhage. In these cases of dyscrasia any slight wound or minor operation may be followed by a fatal hæmorrhage. Fibrin elements are present also in chyle and lymph, which therefore coagulate like blood when withdrawn from their vessels.

Before coagulation the blood consists of:—

Corpuscles }
Plasma } containing the { Corpuscles. Fibrin. Albumen. Water. Salts. }

Coagulation changes it:—

Clot contains the........... { Corpuscles. Fibrin. }

Serum contains the .... { Albumen. Water. Salts. }

In from 12 to 24 hours after death the blood generally coagulates, even in the heart and larger blood-vessels. During life it may coagulate from local arrest or an impediment in the circulation in a particular part. If blood be accidentally extravasated into the connective tissue, the substance of the brain, spinal cord, or serous cavity, or if the circulation in a blood-vessel stops, it coagulates. If a ligature be placed upon an artery (A 1, Fig. 16) the blood coagulates (2), the clot filling the vessel and extending from the ligature up to the nearest arterial branch (3) towards the heart—*i. e.*, the point at which the circulation continues. The end (1) of the ligated vessel towards the periphery completely contracts: When the blood gradually coagulates upon the inner surface of the vessel it forms a partial thrombus (C 6, Fig. 16), through the open center of which it may continue to flow. A transverse section of such thrombus looks like a tube with thick walls (D). The thrombus may become complete, however, if the general circulation becomes so sluggish as to result in an increased deposit of fibrin. The materials of coagulated blood, after solidification, lose their original properties, which cannot be restored, being similar in this respect to the albumen of a hard-boiled egg.

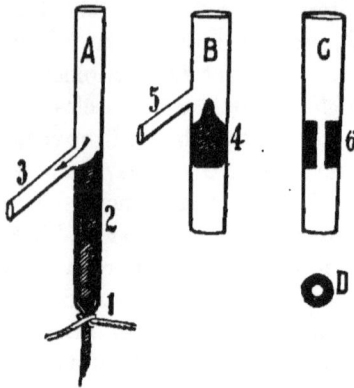

Fig. 16.

Diagram of arteries ligated and clotted.

A 1, Ligation. 2, Clot. 3, Branch through which the blood continues to circulate. B 4, Thrombus. 5, Branch open like 3. C 6, Partial clot. D, Section through a partial clot.

A **Thrombus** is a clot of coagulated blood in a blood-vessel (B 4, Fig. 16) produced by the adhesion of fibrinous molecules one to another. Having strong mutual attraction when in contact, the molecules tend to expel with increasing force the fluid contained, thus forming *fibrin*. If a fibrinous formation take place in the circulatory apparatus, it usually assumes the shape of a conical plug, filling up the lumen of a small or medium-sized vessel towards the still open blood-current by a conical point, and creating thus an obstructing thrombus (4). The thrombus, then, is a foreign body. It may become absorbed, in which event it gradually diminishes in size, disappearing in a few days, or in as many months.

Thrombi differ, however, in structure, the structural peculiarity depending upon the rapidity or slowness of formation. If an artery be ligated and the thrombus form rapidly, it consists of unstriated fibres mingled uniformly with red and white corpuscles regularly dispersed; but if formed slowly, the white corpuscles lay stratified. The red corpuscles, with the fibrinous material of the blood, form the principal mass of a thrombus, which, when the corpuscles part with their coloring matter (which quickly takes place), appears white. If a clot form in the pulmonary veins, or left side of the heart, it may pass either to the brain, at the middle cerebral arteries in the neighborhood of the Sylvian fissure, to one of the kidneys, or it may lodge in the popliteal artery behind the knee; if formed in the venous system. or right side of the heart, it will enter the lung. If the clot be not larger than a blood corpuscle, it may easily pass to a capillary, but if it exceeds this size it will remain and occlude the vessel whose lumen, or size, it cannot pass, thereby producing an *embolism*. The blocking-up of a large vessel brings about an anæmic condition of the organ affected; if the obstruction occurs in capillaries it soon produces excessive congestion, as they fill up from all sides with blood. The pressure may be high enough to produce hæmorrhage by rupture of the vessels, the stagnation preventing nutrition of the part. At the height of typhoid fever the blood is very dark-colored; if coagulated, presenting a small and soft clot. In scarlatinal rheumatism the blood is liable in certain parts to coagulate, so that in severe or fatal cases some of the peripheral arteries and larger veins are generally found to be blocked, the former with emboli, the latter with thrombi. White blood-corpuscles have also a great affinity for each other. During health nothing but the rapidity of the blood-current prevents their adhering to each other and their excessive accumulation.

CHAPTER IX.

THE CIRCULATION.

The digested nutritious substances which pass through the intestinal canal, are absorbed partly by the capillary veins, partly by the lacteal vessels, and carried into the general circulation by two great channels. The venous blood of the intestines, stomach, and spleen is carried to the liver, and from there into the inferior vena cava, which discharges into the right auricle of the heart; while the lacteal vessels carry the chyle into the thoracic duct, which discharges into the left subclavian vein at its intersection with the left internal jugular. In this way the nutritious fluids are mingled with the circulating blood. Arriving at a muscular pump (the heart), the blood is forced through strongly lined arteries, whose walls gradually become thinner as the branches grow smaller. At last it arrives at the delicate network of arterial capillaries, consisting of single membranous tubes, through which the nutritious substances of the blood ooze, and serve as nourishment for tissues. The remainder is carried by the capillary veins to larger and still larger veins, finally emptying into the heart.

Dr. Harvey first described the circulation of the blood in 1628. The manner, however, in which the blood passed from the arteries to the veins still remained shrouded in mystery. Dr. Malpighi, who was born in the same year (1628), afterwards discovered, by the aid of a magnifying glass (microscope), the blood corpuscles, and the passage of the blood from the arteries *through the capillaries* to the veins.

The circulatory apparatus consists of four parts:

Heart.—A hollow muscular organ which propels the blood.

Arteries.—Membranous, elastic, pulsating canals which convey the blood from the heart to all parts of the body.

Capillaries.—A network of minute tubes, which bring the blood in contact with all the tissues.

Veins.—Vessels which receive the blood from arterial capillaries among the tissues and return it to the heart.

The **Heart** is a muscular structure suspended in a closed bag or sac termed *pericardium*. It is divided by a vertical and a transverse partition into four cavities—two auricles (A, B, Fig. 18), and two ventricles (C, D). The shaded portion (3, Fig. 17) is called the right, venous, or pulmonary heart, because it sends the venous blood to the lungs; the unshaded portion is called the left, arterial, or systemic heart, because it sends the arterial blood into the general system. The heart is thus the center of a double circulation—a *lesser* circulation of the blood, from the right half of the heart through the lungs to the left half; and a *greater* circulation, from the left half to the general system, the blood returning to the right half. Its size is about that of the closed fist, the average weight being a b o u t 11 ounces in the adult male, and nine in the female. It is attached to the spinal column, a n d suspended somewhat freely in the cavity of the chest, its great blood-vessels passing through the *mediastinum*. The heart has eight orifices; four in the auricles, of which two are for the venæ cavæ(5 and 6,Fig.18) one for the coronary vein in the right auricle, and one (properly four) for the pulmonary veins (9) in the left; two auriculo-ventricular (1 and 2) between the auricles and ventricles; one aortic (4), and one pulmonary (3). Six of these orifices have valves (the in ferior vena cava, the two auriculo-ventricular, aortic, pulmonary and coronary), which guard these orifices, in order to preclude the regurgitation of the blood.

Fig. 17.

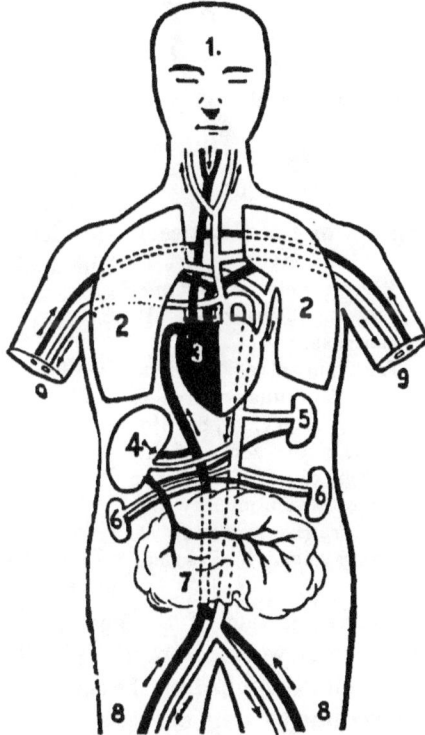

Sketch of the blood circulation.
1. The head. 2, Lung. 3, Heart. 4, Liver. 5, Spleen. 6, Kidney. 7, Intestines, 8, Part of lower extremity. 9, Part of upper extremity.

In the right auricle, the inferior vena cava (5) is guarded by the Eustachian valve, while the superior vena cava (6) has no valve.

Fig. 18.

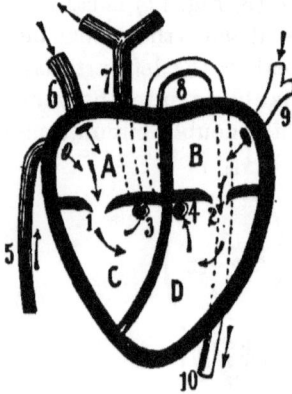

Diagram of the Cavities of the Heart.

A, Right auricle. B, Left auricle. C, Right ventricle. D, Left ventricle.

1, Tricuspid valve. 2, Mitral valve. 3, Semilunar valves of pulmonary artery. 4, Semilunar valves of the aorta. 5, Inferior vena cava. 6, Superior vena cava. 7, Pulmonary artery. 8, Transverse aorta. 9, Pulmonary veins. 10, Descending aorta.

The right auricle has also the valve of *Thebesius*, or coronary valve (not shown in Fig. 18), guarding the coronary vein at a point between the orifice of the inferior vena cava and right auriculo-ventricular orifice. The tricuspid valve is at the auriculo-ventricular orifice (1), between the right auricle and right ventricle. The orifice of the pulmonary artery (3), in the right ventricle, is provided with three semilunar valves. The orifices of the pulmonary veins (9), in the left auricle, have no valves. The mitral valve (2) guards the left auriculo-ventricular orifice. The aortic orifice (4), of the left ventricle, has also three semilunar valves. Thus the two orifices of the vessels (the pulmonary artery and aorta, 3 and 4,) that have to stand the greatest pressure are supplied with three valves. In some cases they have only two valves.

The walls of the ventricles are much thicker than those of the auricles. That of the left ventricle is about four times as thick as that of the right, constituting the greater part of the heart, and indicating the difference in proportion of work accomplished. The blood which nourishes the heart itself is supplied by the coronary arteries—two branches from the aorta immediately above the semilunar valve. This blood, after having discharged its function, is carried to the right auricle by the coronary veins, and has the shortest circulation of any in the body. The two venous blood streams of the venæ cavæ (5 and 6), in pouring into the right auricle, cross each other; i. e., the blood of the two veins passes into the auricle downward and forward, thereby intermixing thoroughly. From this auricle it passes into the right ventricle with sufficient force to turn upon itself; then it is forced upward from that ventricle through the pulmonary artery into the lungs, to give up its carbonic acid and receive oxygen, thereby changing from venous into arterial blood. This arterialized blood passes on through the pulmonary veins to the left auricle, and downward into the left ventricle, where it makes another turn upon itself, and is forced from that ventricle upward into the aorta, and through the entire arterial system. Thus the blood currents are forced through the cavities of the heart with a spiral or twisting motion, caused by the alternate contractions, with relaxations of its walls, termed *cardiac pulsation*. This produces two sounds. The first is distinguished by placing the ear to the chest over the fifth rib and fifth intercostal space, where the sound is loudest; the second, following the first almost immediately, is heard best and most distinctly over the sternum at the level of the third costal cartilage, directly over the semi-lunar valves. The first sound is caused partly by the muscular contraction of the cardiac fibres and tension of the auriculo-ventricular valves, at the time of the ventricular contraction, but mainly by the apex of the heart striking against the inner wall of the chest. The second, or aortic sound, is caused by the sudden closure and tension of the aortic and pulmonary (semi-lunar) valves, lasting half as long as the first. For one-fourth of the whole time of pulsation the heart is at rest. When its muscular fibres are relaxed the blood enters the cavities, causing dilatation, termed *Diastole*. At once the ventricles contract and forcibly expel the blood, termed *Systole*. Simultaneously with the contraction of the ventricles the apex of the heart moves slightly from left to right, and so rotates upon its own axis, as it were. These movements of lateral inclination and rotation are caused by the spiral arrangement of the muscular fibres in the wall of the heart. During the diastole, the transverse diameter of the heart increases and its length diminishes; while during the systole the reverse occurs. Each cardiac pulsation is a double contraction and relaxation, the former commencing at the auricles, followed by the ventricles; then comes a short interval of repose. The auricular contraction is short and feeble, the ventricular being more powerful and lasting longer. These two forms of contraction do not alternate exactly, but are connected and continuous, appearing as one regular motion only, termed *rhythm*, and ranging from 70 to 75 per minute in the adult.

It may at first sight appear absurd to state that the respiratory activity of the lungs and the pulsative activity of the heart are *peristaltic* in their nature; *i. e.*, incited to normal activity by or through local contact, the lungs by contact with carbonic acid and air, and the heart by contact with the carbonic acid of the moving blood—in short, that the primary physical or chemical stimulation for normal action in organs and cells is from local stimulation in all. As soon as we come, in our physiological inquiries, to consider more fully the simple laws of nature, and drop all explanations not based on fact, or not borne out by solid reasoning and deduction, we will begin to truly comprehend the physical laws of the human economy, and not before. No one has ever proved conclusively that either the so-called "vital spot" of the medulla oblongata, the spinal cord, pneumogastric nerve, cervical or cardiac plexus, or ganglia, are, singly or combined, the primary cause of the heart's normal action; and this remark applies just as forcibly to the normal activities of the lungs. To say that all or any of said nervous organs are the *primary* cause of the heart's or lung's activities, is to contravene the law of nature. No physical body possesses the innate power of changing its condition with reference to motion; this is caused only by another body's power of producing motion. Take away from the heart the moving blood as a local stimulus, or the moving carbonic acid and air from the lungs, and they will cease to act, despite the efforts of all the nervous organs mentioned. The carbonic acid stimulus on the right half of the heart produces a reflex impulse through the circulatory centre of the medulla oblongata on the entire heart, similar to the stimulus to the lungs, which produces reflex impulses on the respiratory muscles. It might be objected, in opposition to this view, that if we produce a slight lesion only at the "vital point" of the medulla, the lungs and heart will stop. So they will, but in either case it would be simply an interference with the normal activity of these organs. Every argument must have a resting point, or *point d'appui*—a fact. All that is here claimed is that *the primary cause of every normal activity of any organ or cell within the vital animal economy, other than psychical activity, pertains to sensation.* Now, as no one has yet determined satisfactorily the *primary* cause of the normal activity of the lungs and heart, it is permissible to turn in another direction in our search for that primary cause. This we have done, and have come to the conclusion that it is found in the sensibility of the organs themselves. Again, *to every chemical or physical action*, so far as motion is concerned, *there is an equal and opposite reaction*, which may reasonably be termed *peristaltic* as regards the normal functional activity of an organ or cell in the manifestations of vital animal life. (*Vide* Capillary Action; Tactile Corpuscles; Pacinian Bodies; and External Sensation.) All *normal* actions of the *nervous system* on the heart and lungs—in fact, on any organ of the body (other than nervous centres)—are of indirect origin, and any voluntary or involuntary nervous action, on body or limb, is through reflex only.

The natural physical tendency of any portion of the body is that of contraction, cohesion, and rest; consequently, the normal state of sphincters is contraction and rest, and simple touch, locally applied, increases their contraction; or, when the nervous system acts upon them, through an opposite reflex action, the sphincters dilate. Though this is necessary for the ejection of excrementitious substances and for the comfort of life, the nervous action is in each case reflex as regards the organ acted upon. The sphincter has been selected as a good illustration simply. The stimulus of light on the retina acts similarly in effecting a reflex impulse on the *iris*, though in this case for the purpose of contraction.

It is not, by any means, the object of this argument to belittle the importance of the nervous system; for in the higher animal it performs a higher function than any other organ or system of organs. Through it all the activities of the body are reflected and the organs brought into harmonious and sympathetic relationship. The sole object here sought is to point out that *the nervous centres are not, in a single instance, the primary cause of normal organic actions* outside of themselves; and that these centres have their own local action simply, just as other organs have, brought about by their own irritability and sensitiveness, when the proper stimulus is applied. Thus is revealed the missing link—the phenomenon of low animal organic life existing without a nervous system. All the laws of nature, although not fully understood, are simple. The shortcoming of our reasoning powers is nowhere better illustrated than in the laws governing our own body. For hundreds of years the best scientific minds have been occupied in explaining these laws, and still many of them are not even yet fully understood. This deficiency can lie only in the fact that we do not adhere to the simplicity of the laws of nature and of life, but continually strive to introduce explanations as vague as they are unphilosophical and unreal. Through all the phases of organic nature, we are brought face to face with manifestations of a power above the merely physical and chemical. Try to explain protoplasm and the activity of cells on merely physical or chemical grounds, or on both, and the question as to how the protoplasm and cells of the human economy, or other of the higher forms of animal life, become endowed with a power and activity superior to the alleged causes of their production, must certainly remain unanswered. There is but one solution, and this is based on the admission of a *vital principle*, or *soul*. The author does not deny that physical and chemical forces play their part in the phenomena of organic life, but he strenuously holds that there is a power and activity above and superior to these, and without which growth and nourishment—in a word, that activity so much beyond and distinct from the physical and chemical—cannot be accounted for.

In **Syncope,** or fainting, there is failure of the heart's action, causing a suspension of the circulation and an entire cessation of power in both brain and spinal cord, followed by more or less loss of sensation and muscular movement. Such a condition may also be brought on by some powerful emotion. The reflex actions of the nervous centres on the muscular system, respiratory and circulatory apparatus, are, in such a case, interfered with. The heart is easily affected by sentimental emotions, collectively or separately, such as love, hate, joy, or grief, each of which is capable of interfering with the reflex actions. While the cells of the cerebral convolutions are endowed with the power of receiving impressions through the sensation caused by the soul, and thereby affecting the heart, or any other organ—this, as in the case of an impulse from an external (organ) sensation, can only be through reflex action, in so far as the nerve impulses are concerned; for the cells of the cerebral convolutions have no direct communication with any organ other than nervous centres, so that the stimulus from the vital principle causes an impulse on the cells of said convolutions, and the impulse is forwarded to another nervous centre, where it is reflected. If, for instance, the impulse be reflected from the corpora quadrigemina, it may cause a sense of darkness before the eyes; from the cerebellum, either activity or loss of muscular power; from the medulla oblongata, either increased action or impairment of the respiratory system, circulatory apparatus, or digestive organs; from the spinal cord, generally relaxation of the sphincters.

Arteries distribute the blood outwardly from the heart throughout the entire body. The **Aorta** begins at the heart, and in its course gives off large branches (arteries), which subdivide into smaller ones, finally terminating in capillaries in the tissues. The **Walls of the Arteries** are composed of three coats: an inner, termed *tunica intima,* consisting of thin elastic laminæ lined with a single layer of nucleated squamous cells; a middle, called *tunica media,* consisting principally of non-striated muscular fibres encir cling the vessel transversely; finally, an external coat, the *tunica adventitia*, consisting principally of elastic fibrous connective tissue.

The principal anatomical distinction between the larger and smaller arteries is in the structure of their middle coat. In large arteries this coat consists mainly of thick connective and elastic fibrous tissue, which in the medium sized vessels is less in quantity, while in the smaller this layer consists exclusively of muscular fibres. The large arteries are therefore very elastic, with little contractility, while the smaller are very contractile, but less elastic. The capillaries consist of a layer of elastic connective tissue lined with nucleated squamous cells.

The arteries possess two properties influencing the movement of the blood within; viz., elasticity and contractility. *Elasticity* is the property by which the interrupted or intermittent action of the heart is made equable and continuous, chiefly in the larger trunks; while *contractility* is more concerned in regulating the flow of blood in the smaller vessels.

Arterial Pulsation is produced by the ventricular systole (contraction) of the heart, forcing the blood into the arteries and distending their walls by the pressure exerted, and the subsequent contraction of the arterial walls. There are, therefore, two forces contibuting to produce arterial pulsations—the heart's contraction, and the elasticity and contracting force of the arteries. The arterial distension and elongation increases with age, and the arteries finally become permanently enlarged, especially in their curvatures. During relaxation of the ventricles of the heart, the semilunar valves close, and prevent any movement of the arterial blood backward. The blood is thus forced onward under the pressure of the arterial elasticity and peristaltic action into the smallest capillaries. (*Vide* Capillary Action.) When the arteries have become partially emptied they are again distended by another contraction of the heart, and by this successive expansion and relaxation pulsation is produced. The blood flows much faster through the arteries than through the capillaries and veins. Its greatest velocity is in the aorta nearest the heart. The average velocity in the carotids is about twelve inches per second. As the blood approaches the capillaries the current diminishes in rapidity, owing to the mechanical friction on the arterial walls, also to the fact that the combined calibre of the capillaries largely exceeds that of the single vessel leading to them; so that in the capillaries the velocity is only about one and a half inches per second. No pulsation is perceivable in the capillaries, on account of their small size and plexus formation.

The **Pulse Rate** in the *fœtus in utero* is 142 per minute; in the new-born infant, 132; from 1 to 4 years, 105; at 15, 85; and from 25 to 60, about 70 to 75. The rapidity of the heart's motion is inversely to its power. Thus, a rapid pulse indicates a feeble action of the heart, and a normally full pulse a strong action. In *hydro-cephalus* the pulse may rise to 200 per minute, or in *apoplexy* fall to 30 or even to 20. (For certain abnormally slow pulsations, see Bile.) Sometimes we find a *dicrotic* or double pulse; *i. e.*, two pulsations of the artery for each contraction of the heart. One is caused by the impulse of the heart, the other by the unequal elasticity and distension of the artery. In *typhoid* disease, this condition of pulse is often present. Again, a dicrotic pulse does not always denote a diseased condition, for it sometimes exists in the healthy individual—in such a slight degree, however, that it cannot be detected with the finger. With the *sphygmograph*, or automatic register, it is observable.

Veins are tubular canals which convey the blood from the capillaries of the tissues toward the heart. They are like arteries in structure, having three coats, an inner, middle, and exterior. These coats have the same names as those of the arteries. They differ from them, however, in containing fewer muscular and elastic fibres, but more connective and white fibrous tissues; showing that veins have more distensibility but less resistance than arteries. The veins, like the arteries, are found in all tissues (except in muscular fibres, the crystalline lens, vitreous humor, superficial layer of the cuticle, hair, and nails). The veins begin and form continuous channels from the minute capillary plexuses (Fig. 20). They may be divided (excepting certain venous sinuses within the skull) into two sets—*superficial* and *deep*. The latter generally accompany the arteries in one sheath of fibrous and elastic tissue. The smaller arteries are accompanied by a pair of veins, one on each side, termed *venæ comites*. Immediately beneath the integument, the superficial veins discharge their contents into the deep ones, and all finally unite in two large trunks, termed *superior* and *inferior venæ cavæ* (5 and 6, Fig. 18), which open into the right auricle of the heart. The arterial portions of the capillaries give off and supply blood to the tissues, while the venous portion of the capillaries carry the blood from them, whence it flows into larger veins.

The pulmonary, portal, and umbilical veins act in the capacity of arteries. Veins through which the blood flows contrary to the natural law of gravity are supplied with a great number of check-valves formed by a doubling of their internal lining membrane, pocket-like folds, or pouches (Fig. 19). The other veins, such as the superior vena cava, pulmonary, renal, portal, and hepatic, have no such valves. The venous valves permit the blood to flow readily toward the heart, but prevent it from returning toward the periphery. The flow of blood through the venous system depends on the favoring conditions of the veins and certain forces: 1. The walls of veins offer no resistance to the blood, as the lumen of veins increases as they approach the heart. 2. The alternate contraction and relaxation of the voluntary muscles exerts some pressure on the veins between the muscles. 3. The respiratory distension of the chest and descent of the diaphragm impel an onward movement. 4. The suction caused by the expansion of the right auricle of the heart. 5. Capillary peristaltic action. (*Vide* Capillary Action.)

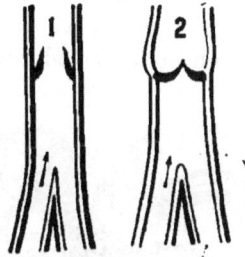

Fig. 19.

Diagram of the Valves in Veins.

1, Valve open. 2, Valve closed.

In the venous circulation of the stomach, spleen, pancreas, and intestines the blood does not enter the vena cava directly. The veins of these organs unite in one large vessel, the *vena porta*, which enters the liver, where it branches out like an artery, subdividing into a capillary network that permeates the whole mass of the liver, supplying considerable material from which the bile is secreted, and to which the arterial blood of the hepatic artery also contributes. Thus, the blood flowing through the portal vein passes through two sets of capillaries in the interval between leaving the aorta and entering the vena cava.

The movement of the blood in the veins is only about one-third as fast as in the arteries. The capacity of the entire venous system is about two and a half times as great as that of the arterial.

The **Capillary Circulation** is the most important feature of the circulatory movement. Through the capillaries a free interchange between the blood and tissues takes place. The capillaries (*capillus*, a hair,) are minute blood-vessels, forming the connection between the terminal branches of the arteries and the commencement of the veins. In size they average about 1-3000th of an inch

7

in diameter, and their arrangement varies in the different tissues. They pursue no definite course. In the muscular system they are somewhat parallel with each other; in the skin and intestines they form loops; around fat-cells they are arranged spherically (Fig. 20). When blood-vessels approach the capillary system they gradually diminish in size, first losing their external coat, the middle coat at the same time becoming thinner and gradually disappearing altogether. The inner coat alone remains, forming the capillary, a thin elastic fibrous connective tissue layer, lined with nucleated squamous cells. The walls of the capillaries are, therefore, soft and yielding, so that osmosis of liquids and gases can readily take place. The arterial as well as the venous capillaries form networks by free inosculation, termed *capillary plexuses*. The largest capillaries are found in the secreting glands and osseous tissues; the smallest in the muscles, nerves, and retina. The smallest are about the size of the minutest red blood corpuscles. In each organ and tissue the inosculations of the capillaries differ; in the organs of secretion and absorption they are the most numerous.

Fig. 20.

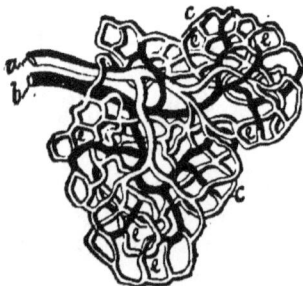

Diagram of the Capillary Circulation.

a, Artery. *b*, Vein. *c c*, Arterial capillaries. *e e*, Venous capillaries.

The combined calibre of the arterial and venous capillaries increases in diameter so enormously that it is about 800 times as great as that of the parent trunk; consequently, when those circulatory vessels have divided and subdivided into capillary networks, the heart's action on those vessels becomes indistinct, and no pulsation in the capillaries is perceptible.

The **Capillary Circulation is Peristaltic**; but, on account of the minuteness of the diameter of the capillary, it is difficult to illustrate the rapid alternation of contraction and relaxation which takes place. From a likeness of effect we argue a likeness of cause. Hence, the manner in which fluid matter is propelled through the capillaries cannot be contrary to that in other tubes of the animal economy with a similar function. In the chapter on "The Senses" it is stated that sense-organs are endowed with the power of reacting to appropriate stimuli. The presence of blood in the capillary causes normal peristaltic action, similar to the action of the ingesta in the alimentary canal, blood in the heart and arteries, lymph in the lymphatics, and urine in the ureters. When the position of the body is reversed (the head downward and the feet upward) the contents of the stomach continue to pass into the intestine by virtue of the peristaltic action of the œsophagus, which the closure of the cardiac orifice alone cannot prevent. Again, in this position of the body the urine persists in flowing to the bladder, the intestinal contents towards the anus, and the blood to the feet, though not with the same force, as the position of the body is unnatural. That the circulation in the animal capillaries is peristaltic may be further illustrated by the fact that when an object gives off matter it becomes smaller, while the object attracting or absorbing it becomes larger, for no two particles can occupy the same space at the same time. The capillary is fed continuously, and the blood in the capillary cannot recede in the normal condition on account of the arterial force on the blood and the capillary peristaltic action forward. The capillary gives up nutritious substances to the surrounding absorbing cells, which in their activity must alternate in dilating (taking up matter) and contracting (giving off matter to the tissues), so that the dilation and contraction of the cells in the delicate wall of the capillary produce the peristaltic action in the animal capillary, caused by the presence of blood. Experience illustrates these facts, that when the contractility of the arterial portion of the capillaries is relaxed (dilated) in any part of the body, for a longer or shorter period of time, an *inflammation* of the part results correspondingly. On the other hand, when the venous portion of the capillaries abnormally contract the blood accumulates, causing inflammation, which is accompanied not only by an increased quantity of blood in the capillary vessels, but also by an increased temperature.

The following points may be considered in relation to the inadequacy of the heart's action to the capillary blood propulsion: 1. A slight change (peristalsis) is noticeable in the capillary calibre in the frog's foot. 2. Secretions, such as sweat, may occur after the heart ceases to beat. 3. In certain animals of the cold-blooded type the capillary circulation has been noticed to continue for some time after the heart was excised. 4. Animals, even the fœtus, have been produced without a heart, with all the other organs well developed. The illustrations of the reflex requisite for capillary action through nerve-fibres and nervous centres are similar to those given in the chapters on the "Mucous Membrane", "Secretion, and Absorbents."

Capillary Action.—The so-called *capillary action* of plants and of dead matter, be it organized or not, cannot properly apply to capillary action in the vital animal economy; for in the former there is no sensation and no peristalsis, but simply molecular and atomic attraction, cohesion, and adhesion. Thus a piece of soft wood, a sponge, a towel, a piece of sugar, etc., become wet throughout when a small part of them is dipped in water.

Organs and secreting glands, during their activity, become excited, and more blood passes through them; but during activity the nourishing blood for the tissue of the organ or secreting gland retains its arterial color even to the veins; hence, the nourishment of the tissues of the organ or gland takes place only during repose, when the arterial blood becomes venous, while in the muscular system the greater the activity the darker the venous blood. In paralysis, in complete muscular repose, or complete etherization, however, the venous blood of the part affected has the color of the arterial. As the blood yields to each organ and tissue the material required for either nutrition or secretion, the venous blood returning from each organ or tissue is consequently altered in composition. In fact, the venous blood being no longer fit for the same purpose, it is returned to the heart to become there intermixed, and, not only in the heart, but also in the lungs, to receive new material. (*Vide* chapter on Respiration.)

CHAPTER X.

RESPIRATION.

Respiration involves inspiration and expiration. In inspiration about one-fifth of the amount of oxygen inhaled is absorbed by the body, while in expiration the carbonic acid gas—or, as more recently stated, *carbon dioxide*—is eliminated.

The **Trachea** is a large tube formed of thick cartilaginous rings connected by elastic tissue. The rings serve to prevent the collapse of the tube (6, Fig. 21). Before entering the lungs the trachea bifurcates, forming two bronchi, the excretory ducts of the lungs. After entering the lungs each bronchus divides and subdivides into smaller tubes (bronchial tubes) until their diameter is less than one millimetre (0.039 inch); their further subdivisions are termed *bronchioli*, which terminate in the *alveoli*. (*Vide* Fig. 22.)

The **Lungs** (1, 2, Fig. 21) are the essential organs of respiration, the right lung being broader, but about an inch shorter than the left. This is because the diaphragm is higher (to accommodate the liver) on the right side than on the left. The weight of the right lung is about 22 ounces, while the left, which is more elongated in form (by reason of the heart lying partly inferior and anterior to it), weighs about 20 ounces. The tissue of the lung is of a pinkish slate color, light, porous, and spongy, floating if placed in water, or crepitating if manipulated, on account of the presence of air in its alveoli. It is also highly elastic, which accounts for its collapse when removed from the closed cavity of the thorax, or when air enters the pleural cavity. The *parenchyma* of the lung consists of minute lobules, blood-vessels, lymphatics, and nerve-fibrillæ. The greater part of the lung is composed of lobules, each formed by a cluster of *alveoli*—also termed *air-cells* or *air-vesicles*—and held in position by minute, delicate, and very elastic tissue; indeed, each lobule is, in function, a lung in itself, so that if the major portion of the lobules are destroyed, as in tuberculosis, respiration still goes on, until there are insufficient lobules left to adequately carry on the respiratory function; *i. e.*, to supply the blood with oxygen and to remove the carbonic acid. A great many lobules are aggregated by connective tissue to form large lobes, two of which form the left lung and three the right.

Physiologically, the respiratory organ (lung) consists essentially of moist and permeable membranes, with blood on one side and air on the other; yet the blood and air do not come in direct contact, but are absorbed and transude through the membrane between. All the alveoli together form in the adult lungs a respiratory surface about 128 square metres in extent. The blood, therefore, in being distributed over so large a surface, is more completely arterialized. The exchange of carbonic acid gas for oxygen is effected very rapidly. Gases possess almost perfect freedom of molecular motion.

Fig. 21.

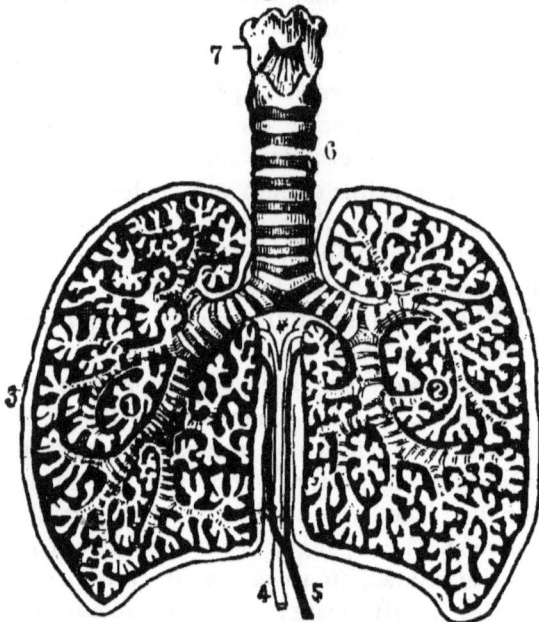

Diagram of the Lungs.

1, Right lung. 2, Left lung. 3, Pleura. 4, Pulmonary artery .5, Pulmonary vein. 6, Trachea. 7, Cricoid and Thyroid cartilages.

Owing to their unlimited tendency to flow, both liquids and gases are called fluids. It is natural for each substance in the mixture, if unobstructed, to flow from places where it exists in greater quantity to places where it is less abundant, and, according to two of the four laws of capillary action, the ascension and depression of a liquid not only varies with the nature of the substance, but also inversely in conformity with the temperature.

The **Nerve-fibres** of the lungs are derived from the anterior and posterior pulmonary plexuses, composed of filaments from the pneumogastric and sympathetic nerves. Some efferent respiratory nerve-fibres, though arising in the respiratory centre in the medulla oblongata, pass down the spinal cord and emerge from between the third, fourth, and fifth cervical vertebræ, collecting on each side of the cord into a nerve, called the *phrenic*, which supplies the half of the diaphragm on the corresponding side of the body.

The **Phrenic, or Diaphragmatic Nerve,** arises from the union of the branches of the third, fourth, and fifth cervical pairs. It passes between the clavicle and subclavian artery, and descends by the pericardium to the diaphragm. If the phrenic nerve alone be divided outside of the spine, the channel for reflex impulses to the diaphragm being cut off, the latter is said to be paralyzed, and respiration is observable solely in the rising and falling of the ribs. A division or compression of the intercostal nerves only, paralyzes the intercostal muscles then respiratory movements are performed solely by the diaphragm. Division of the spinal cord above the third cervical vertebra interferes with the transmission of the reflex impulses through the phrenic and intercostal nerves, resulting in paralysis of the movements of the chest and abdomen, though the glottis, mouth, and nostrils may continue to act for a short time; the final efforts to respire are manifested in gasping and convulsive movements of the nostrils, for the reflex impulses from the medulla oblongata to these muscles still continues. If the medulla oblongata be injured, the power to reflex an impulse to breathe is at once taken away, and respiration ceases.

That there are other nerve-fibres connected with the respiratory centre, besides those of the pneumogastric and phrenic, there can be no doubt. If, for instance, the nerve-fibres of the skin and chest are stimulated by an application of cold water, a gasping inspiration ensues. The nerves of the soles of the feet, or those of the skin of the sides of the body, if stimulated by tickling, causes laughter on account of a series of peculiar spasms of the diaphragm, abdominal, and facial muscles. A funny sight, the recollection of or present impression on hearing a humorous speech, causes similar spasms.

The **Respiratory Centre, or Vital Point,** at which the rhythmic movements of the respiratory activities are regulated (reflexly), is at the floor of the fourth ventricle in the medulla oblongata, its *afferent* (sensory) nerve-fibres being those of the *vagi*, and the *efferent* (motor) those of the phrenic and intercostal nerves.

Lymphatics.—The lung is well supplied with lymphatic capillaries, which take up not only the rudimentary matter from the lung tissue, but also the minute solid particles entering the alveoli with air, as noticed in the lymphatic-vessels of coal-miners.

Blood of the Lungs.—The lungs are supplied with two kinds of blood, one for nutrition, the other for purification and stimulating respiration. Consequently, the capillaries are from two sources: from the bronchial arteries and from the pulmonary. The *bronchial arteries* (from the aorta) furnish blood to the parenchyma of the lungs for nutrition; this blood is returned through the bronchial veins to the superior vena cava. The *pulmonary arteries* (from the heart) divide in the lungs into networks of capillaries, which run along on the inner surface of the alveoli. These capillaries unite to form larger venous branches, which again unite to form two large pulmonary veins from each lung, and discharge their contents into the left auricle of the heart.

The **Pulmonary Artery** (4, Fig. 21; and E, Fig. 22) conveys the venous blood from the right ventricle of the heart to the lungs,

Fig. 22.

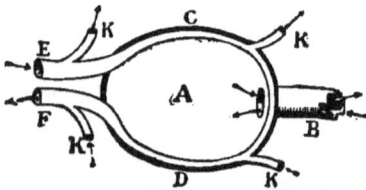

Diagram indicating Alveolus and Capillary Blood-vessel.

A, Alveolus. B, Bronchiolus. C, Pulmonary artery portion of the capillary. D, Pulmonary vein portion of the capillary. E, Pulmonary artery. F, Pulmonary vein. K, K, K, K, Capillaries of the adjoining alveoli.

where it divides into minute capillaries, which pass along the walls of the alveoli, adhering to them on the inner side by a small surface only, leaving the greater circumference of the vessels exposed to the air. This allows the blood in the capillary (C, D, Fig. 22) to come almost in direct contact with the air in the alveolus (A). The delicate membrane of the pulmonary capillary through which osmosis, or interchange of the gases of the blood, takes place, is less than 1-20,000th of an inch in thickness; the diameter of the capillary (C, D,) being from 1-2000th to 1-1500th of an inch. Each **Alveolus** (A) is a small recess at the termination of a bronchiolus (B), and is from 1-200th to 1-100th of an inch in diameter, and separated from the adjoining alveoli by thin walls or septa; consequently, they do not intercommunicate, but are held together by connective tissue. The diameter of the bronchiolus (B) is from 1-500th to 1-400th of an inch. It is estimated that the lungs of an adult contain about three millions of alveoli, the walls of which are lined with a single layer of squamous epithelial cells. The walls contain the blood and lymphatic capillary vessels and nerve-fibrillæ. The network of capillary blood-vessels communicates from cell to cell, and the blood in the vessels remains exposed to the air in the alveolus (air-cell) about a second and a half to eliminate the carbonic acid, and absorb oxygen. At each inspiration, air enters the alveolus (A), through the bronchiolus (B); and the oxygen is absorbed by the hæmoglobin in the capillary portion D. It cannot enter the portion indicated by the letter C, as this portion is filled with venous blood saturated with carbonic acid, the gas of which is just escaping. The carbonic acid issues from the blood at C, and the blood on arriving at D is capable of absorbing oxygen. That the oxygen is absorbed at D, but not the carbonic acid gas, although intermixed more or less in the alveolus A, may be explained in two ways: 1. The powerful attraction of the hæmoglobin, with its iron, for oxygen, and its want of attraction for carbonic acid gas. 2. The heated carbonic acid gas has a powerful expansive force towards a free surface, especially towards the colder gas (the atmosphere) externally. This explains why respiration in an atmosphere of higher temperature than that of the blood is more difficult and hurried. In this case, the carbonic acid gas is imperfectly carried off from the lungs, and the skin assumes greater activity (sweating) to compensate for this deficiency. The lungs manifest a wise arrangement in the conveyance of venous blood by an artery (pulmonary), as an arterial vessel only can withstand the pressure of the ventricle of the heart; so, also, in the conveyance of arterial blood by veins (pulmonary), which easily relieve the lungs of their blood.

The **Four Pulmonary Veins** return the aerated blood (arterial) to the left auricle of the heart; consequently, it is owing to the process of respiration that the dark venous blood is changed into a bright arterial red. This blood, in passing through the capillary vessels of the body, supplies the tissues with oxygen and with nutritive substances, and during this process gradually becomes reconverted into venous blood and returned to the heart, which sends it through the pulmonary artery to the lungs, in order to again receive a fresh supply of oxygen. The local contact of the gases (air and carbonic acid gas) and the organic matter in the blood in passing through the lungs furnishes the stimulus for continual activity of these organs. The carbonic acid and the organic matter are, no doubt, the irritating substances—the oxygen the alleviating (satisfying) element.

The respiratory centre in the medulla oblongata acts by reflex; just as do all nervous centres by which impulses on organs are reflected. Each centre is connected, either directly or indirectly, with the organ by afferent (sensory) and efferent (motor) nerve-fibres. This is the best proof that the sensation of any organ lies primarily in the organ itself, and not in the nervous reflex centre. Were its seat in the latter, then the afferent fibres would be superfluous, and the efferent sufficient. Consequently, the stimulation and sensation for respiration lie in the lungs, and not in the so-called *vital spot* of the medulla oblongata, this being simply the spot, or centre, where respiratory reflex actions are accomplished.

The **Hæmoglobin** of blood-corpuscles is formed completely in the capillaries of the lungs. It is allied to the albuminoids (proteids), and forms the chief constituent of the red corpuscles. It contains, in addition to carbon, hydrogen, oxygen, nitrogen, and sulphur, about 4 parts of iron in 1000 of blood. The absorption of oxygen in the lungs depends mainly on the molecular attraction inherent in the molecules of the hæmoglobin. When iron is deficient in the blood, the hæmoglobin is defectively developed, and the supply of oxygen and nutritive substances to healthy tissue-changes in the brain, nerves, muscles, etc., becomes lessened; hence, the defective vitality, general or local, during such deficiency. When arterial blood in the capillary circulation comes in contact with the tissues, the red blood-corpuscles furnish them with nourishment by giving up the greater part of their oxygen, iron, hæmoglobin, and the potassium salts. The red corpuscles become anæmic when parting with these constituents. Some of the corpuscles then become disorganized and broken up; though the greater number return to the heart and lungs for recuperation, or for another supply of the above-mentioned constituents. They possess, however, no properties in common with those of the carbonic acid, except that both (carbonic acid and oxygen) pass through the same channels. The hæmoglobin of the corpuscles entering the liver furnishes one of the basic constituents essential to the secretion of bile.

The **Hæmatin** of the blood is of a dark-brown color, and results from the decomposition of hæmoglobin. The term "hæmatin" is derived from the Greek word meaning blood.

The movements of the lungs, glottis, chest, and abdomen correspond with each other in point of time and intensity, so that the same reflex by nervous centre influence that expands the chest for inhalation opens the glottis to admit the air. Only about 1-5th of the air in the lungs is renewed at each inspiration. The number of respirations in the adult averages about 18 per minute; in youth, 20; in childhood, 26; in infancy, 40. During inspiration a rustling sound, termed *vesicular murmur* (crepitus), is produced by the alveoli, and is more distinct in children than in adults.

Like pulsation, respiration is involuntary and purely through reflex. By the force of the will we may partially change or arrest the rhythm of respiration, though only temporarily, or until the intensified stimulation on the lungs, caused by the accumulation of carbonic acid elements in the capillaries of the lungs, forces respiration. This is why a person can remain under water for a short time only.

The expansive force of the lungs depending on the thoracic extension, they collapse with the thoracic contraction. During inspiration the diaphragm descends (contracts) and the intercostal muscles raise the ribs and sternum, enlarging thereby the cavity of the chest. In children and in the adult male the respiratory movements are most noticeable in the abdomen, but in the adult female the upper part of the chest is put in more active motion. During the inflation of the lungs with air (inspiration) the diaphragm contracts, while during expiration this action is reversed. In disease or injury of the lower cervical or upper dorsal region of the spinal cord, the transmission of the reflex efferent impulse being interfered with, the intercostal muscles are paralyzed and the thoracic respiratory movements arrested; then the movements of the abdominal respiratory muscles will be increased. In peritonitis, or other forms of abdominal inflammation, the movements of the abdominal muscles and the diaphragm are restrained on account of the pain present, and the thoracic respiratory movements are increased. The extension and distension of the many organs during inspiration is the principal organic force which, in reaction, produces expiration. It may be here noted that the contraction of the intercostal muscles is not inspiratory, but expiratory—a function opposing that of the diaphragm; i. e., the contraction of the latter favors inspiration; of the former, expiration. An adult requires about 350 cubic feet of air per day. When the inhaled air contains less than 10 per cent of oxygen, asphyxia and then death ensue. In such cases the air is already more or less saturated with carbonic acid gas, so that the power of its elimination from the lungs is diminishe l and the inspiratory stimulus increased; but, as little or no oxygen enters, the increased stimulation is not relieved (satisfied), and the cells of the alveoli of the lungs become so exhausted that little or no stimulus is forwarded to the reflex centre of the medulla oblongata, and no reflex impulse is conveyed to the respiratory muscles for expiratory movements; and sooner or later not only the lungs, but all the muscles of the body are affected by the carbonic acid stimulation.

This results in greater or less convulsive movements, succeeded by tranquillity. This condition is due to exhaustion (asphyxia).

The changes effected in inspired and expired air are as follows:

Composition of inspired air $\begin{cases} \text{Nitrogen.......... 79.02 by volume.} \\ \text{Oxygen........... 20.94 " "} \\ \text{Carbonic acid..... .04 " "} \\ \hline \text{100.00} \end{cases}$

Composition of expired air $\begin{cases} \text{Nitrogen.......... 78.25 by volume.} \\ \text{Oxygen........... 16.25 " "} \\ \text{Carbonic acid..... 5.50 " "} \\ \hline \text{100.00} \end{cases}$

Should air enter the pleural cavity by an opening through the skin and thoracic wall, or from within the lung or respiratory passages, the lung collapses (pneumothorax), and is no longer capable of distension. If this happens to both lung cavities, death results. In case, therefore, it becomes necessary to open the pleural cavity care must be taken that no air enters and that the external opening be well protected until healed. Some water passes from the blood by respiration and perspiration, the kidneys not being able to eliminate all that is required to be removed. From the lungs about 10 to 18 ounces of watery vapor are exhaled in 24 hours, being increased or diminished according to the temperature and moisture of the atmosphere and body. A person with a dry and hot skin respires very frequently. When, however, thorough perspiration takes place, respiration becomes less frequent. In the canine family the tongue is often protruded, and respiration doubled and trebled in frequency by the increased stimulation of the carbonic acid caused by increased muscular activity or by exposure to external heat. The dog's integument contains but few perspiratory glands, and even these few empty into hair follicles, instead of discharging directly on the surface of the skin.

Besides carbonic acid gas and water, a small quantity of ammonia and organic matter is emitted by the lungs and skin, which gives a faint odor to the breath and perspiration; this is very noticeable in a crowded and confined space, and causes the atmosphere to become oppressive. The amount of ammonia and of organic matter exhaled has not yet been definitely ascertained.

Carbonic Acid Elements exist in *combination with bases* in blood-plasma, teeth, and bones. The carbonic acid *gas*, as such, does not exist in the blood in the form as excreted any more than fibrin, urine, tears, perspiration exist as such in the blood, but the gas is completely formed (secreted), as the urine, tears, and perspiration are, at the organ of exit; *i. e.*, said gas is secreted from the *material* existing in venous blood, by the cells of the capillaries in the walls of the alveoli of the lungs, sudoriparous ducts of the skin, and uriniferous tubules of the kidneys. 'The cells of organs have their own way of doing things, and very likely will keep their secret. No one has ever been able to separate or produce by art or science a genuine normal urine, tears, perspiration, gastric juice, or any other secretion from blood, as the living cells of the body do. As well try to separate or produce the genuine normal fragrance of a growing plant or flower; the product may be ever so deceiving, but it is not genuine. Carbonic acid has the formula $C O_2$; *i. e.*, one atom of carbon and two atoms of oxygen; yet it cannot be denied that when this *gas* issues from the living animal body it also contains ammonia, nitrogen, and organic matter.

The alkalinity of the blood, especially the carbonate salts, coming in contact with the acidity (a form of lactic acid) of the tissues, may effect the combination of atoms essential to the formation of carbonic acid, but the formation of gas is not complete until the material has passed through the cells of the organs of exit. An aqueous solution of carbonic acid reddens blue litmus paper. It may be said that whenever alkali and acid, with moisture, come together, carbonic acid gas immediately results; but in no instance can it be shown that the action of chemical elements and the results of chemical reactions are the same immediately within the tissues, including the blood of the living animal economy, as when mixed chemically pure and outside of the body. The chemical constituents of the tissues are all known, and yet we are at a loss to explain on chemical or physical grounds that final transformation of food by which it becomes a part of the *living* animal body; neither can we, by chemically combining the fifteen elements which are found in the human body, produce an *organic vital being*. This will be rendered clear, if we but consider that a thing cannot give what it has not. A combination of any number of dead inorganic elements can never produce a vital organized being, according to the philosophical principle, that the effect cannot be above or greater than its cause. Vitality is above chemism and superior to all physical and mechanical energy. The task of science is to strive after truth, without having any secondary object in view. No doubt chemism may explain the physical activity of organs; but it cannot solve the problem of animal vitality any more than it can the merely physical or mechanical forces seen in *nature* about us.

In every bodily action food is required to supply the wasted ener-gy. This food is not only the matter taken into the system and assimilated, but anything which goes to supply the want of any particular organ or tissue; be it an acid, alkali, or salt, etc. Thus an inactive animal body, suddenly becoming active, requires an increased diet.

It may be noticed that the results of experiments on mamma-lians are not mentioned here. This is because they are so conflict-ing; one claiming that the amount of carbonic acid exhaled is in-creased during cold weather—i. e., in December, January, and Febru-ary—and decreased in June, July, and August, respectively; another, that it is greater in warm weather; or in a moist than in a dry atmosphere, or vice versa. All experiments have been conducted to ascertain the quantity, quality, and possible action of said gas within the living body, though when outside of its environment—even when taken direct from the alveoli of the lungs; but it cannot be true that this external gas is the same as the materials when in the venous blood, and for the reasons already mentioned. If the air-passages be closed completely the respiratory efforts become deep, labored, energetic, and rapid, and the circulation increased (asphyxia). This also indicates that the primary stimulus for inspiration and circulation is the carbonic acid elements, within the blood, on the lungs and heart; and not the oxygen. The use of oxygen is, of course, to satisfy or moderate overstimulation. The cause of the revolution of a rapidly turning wheel is the power that turns it, and not the brake that modifies the motion. The respiratory and circulatory reflex centres of the medulla oblongata are very close neighbors—possibly one; so that a stimulus from the mind, or stimulation from the carbonic acid in the blood on the heart and lungs, is reflected by the respiratory or circulatory centre through efferent impulses on the heart and lungs, correlating the activity of these organs. Cold-blooded animals exhale very little carbonic acid gas, comparatively, indicating that their tissue waste is not so rapid and the material for carbonic acid gas in their cir-culating fluid is less; therefore, the primary stimuli for circulation and respiration are not so marked as in the warm-blooded animals. Plants absorb carbonic acid gas by their leaves, peripheral cells, and roots. The leaves act as respiratory organs, absorbing carbonic acid gas (as may be easily proved by the celebrated experiment of Priestly), and giving off oxygen, which, in the plants, is derived by transformation of water, carbonic acid gas, and of various salts.

The stimulus on the lungs for inspiration is the carbonic acid elements acting on the epithelial cells in the walls of the alvioli, and the stimulus for expiration is the satisfying effect of the atmospheric oxygen on said cells. An excess of carbonic acid in the system causes depression of spirits. This may be relieved by tea or coffee, which produces an increased elimination of this acid, at least greater than that supplied by the food. Carbonic acid eliminated from the animal organism, and again introduced into the body, acts as a poison, similar to that produced by chemicals and introduced in the gaseous form alone; while the chemical gas moderately introduced with water (soda water) into the stomach diminishes thirst, acts as a diuretic, lessens morbid heat and irritability of the stomach, and is slightly anti-emetic. Respiration in an atmosphere deficient in oxygen becomes hurried from overstimulation of the carbonic acid, and, if continued, respiration may soon cease altogether.

It may be noticed here that this theory of **external organic sensation** contravenes the teaching of the present day touching the primary cause of the normal activity of organs in continuous action, such as the lungs, heart, capillaries, kidneys, and skin. (See Respiratory Activity of the Lungs, and Pulsative Activity of the Heart.) The venous blood-plasma being saturated with the elements of carbonic acid, and the ready-formed gas issuing from the arterial portion of the pulmonary capillary (C, Fig. 22), the oxygen has no chance to enter there, though it does enter at the venous portion of the pulmonary capillary (D). The epithelial cells of these vessels, by absorbing and giving off gases, dilate and contract in the walls of the alveoli, especially in the middle portion of the capillaries, producing peristaltic action of said walls; consequently, it aids the movements of the blood within the capillaries, the carbonic acid elements being loosely contained in the blood-plasma. (*Vide* Capillary Action in chapter on Circulation.) The blood-corpuscles have little or no attraction for said elements, or already formed gas, but a most powerful attraction for molecules of oxygen gas. The oxygen itself has decided magnetic properties (paramagnetic), like those of iron; but carbonic acid, water, and nitrogen have decidedly repellent properties (diamagnetic). This offers an explanation as to how both gases, though more or less mixed in the cavities of the alveoli, pass—the one inward, the other outward (A, Fig. 22). When two gases are placed in contact it is impossible to prevent them from mixing, for every gas diffuses into every other. They easily separate, however, and a lighter may rest upon a heavier when one is of higher temperature than the other, as oil rests upon water. The passage of both gases, through the activity of the epithelial cells in the membrane of the capillary, is assisted by the law of diffusion called *osmose*—the oxygen being favored by the powerful attraction of the molecules and the incessant motion of the red blood-corpuscles, the carbonic gas escaping by the expansive force of heat. The pressure of a given body of gas, the volume of which is kept constant, is in proportion to its absolute temperature. In the lungs the formation and pressure of the carbonic acid gas (to escape) are also assisted by the pressure of the blood in the pulmonary artery. The activity of the chest wall, respiratory muscles, diaphragm, and glottis is through reflex, the lungs being stimulated by the carbonic acid for expansion, by oxygen for satisfaction (contraction), each stimulus forwarding an impulse through afferent nerve-fibres of the vagi, respectively, to the respiratory centre in the medulla oblongata, where said impulses are reflected and forwarded by efferent fibres through the phrenic and intercostal nerves, calling into activity all respiratory organs correspondingly, just as the stimulation of normal food on the stomach prompts the emptying of the gall-bladder. An abnormally high temperature of the carbonic acid elements in the blood stimulates the epithelial cells in the walls of the alveoli of the lungs and mucous membrane of the cavities of the heart to increased activity, and consequently to hurried respiration and pulsation. This explains why respiration and pulsation proceed normally in unison.

The absorption of oxygen and the elimination of carbonic acid gas are also effected, more or less, by the activity of the cells of the skin. Covering the entire cuticle with an impermeable substance, such as glue or varnish, causes death. It is like closing the principal air channels, mouth and nose, though the process is slower. This also indicates that it is the local stimulation by the carbonic acid from within, and oxygen from without, on the epithelial cells in the walls of the capillaries of the ducts of the sudoriparous glands that excites activity. Some of the acid gas is also discharged with the urine. That eliminated by the skin and urine amounts to about four and a half per cent of the amount simultaneously exhaled by the lungs. This furnishes another evidence that the continuous stimulus for respiration is the carbonic acid within the blood acting on the lungs; for every organ by which this gas is eliminated is one of incessant activity, as the lung, skin, and kidney. We notice that the elements of carbonic acid, while normally in the system, are of the most vital importance. That the ready formed (exhaled) carbonic acid gas, as such, exists already under the same condition in the venous blood, must be very doubtful, for when it issues from the body and is injected again into the venous blood it kills the animal. Again, that said gas cannot exist in the blood may be illustrated by filling a bottle entirely full of water and introducing a suitable capsule filled with a small quantity of tartaric acid and bi-carbonate of soda. Close the bottle with the palm of the hand, or otherwise, so that it remains absolutely full. As long as the capsule is not dissolved, no action is noticeable in the bottle, but as soon as it dissolves and the atoms of acid and soda become active the bottle bursts, there being no space for the carbonic acid gas which is evolved. Now, if carbonic acid gas in the animal tissues or venous blood existed in the same form as that evolved by the organs of exit, the veins would burst, especially under the pressure of an abnormal increase of this gas; but the fact is that they do not even dilate. Artificial respiratory excitants increase the exhalation; non-excitants diminish it. Albumen, as food, is one of the least respiratory excitants. Temporary cessation of the menses, due to pregnancy or otherwise, increases the exhalation of carbonic acid; so does muscular activity. The carbonic acid gas exhalation is diminished during sleep. Heat being the most general molecular stimulus, a warm atmosphere increases the activity of the cells of the skin, just as a sudorific remedy when taken into the body produces increased perspiration, more carbonic acid gas and heat is given off, and the temperature of the system lowered. Any remedy stimulating the cells of the uriniferous tubules and Bowman's capsules of the Malpighian bodies of the kidneys, acts similarly to increased secretion of the solids of the urine, just as the remedy stimulating the epithelial cells of the walls of the arterial capillaries within said bodies (possibly assisted by the arterial blood pressure) increases the secretion of the watery portion of the urine. Since irritability (sensitiveness) of an organ or tissue is the property of becoming excited to activity when acted upon by the appropriate stimulus, so force causes a body to *move* as well as to come to *rest*. The channels of the blood-circulation serve two opposite purposes in taking in and giving off the elements of oxygen and carbonic acid. The oxygen enters the body through the lungs, skin, and alimentary canal; the carbonic acid leaves it by expiration, perspiration, urine, and the feces.

8

Carbonic acid gas is about one and one-half times heavier than air at the same temperature; but when heated above that of the surrounding air—as when generated by fire, or when escaping with the warm breath of animals—it is lighter, and, consequently, rises.

The human body is composed of about fifteen chemical elements; and, coming in contact with each other in the vital economy, their antagonism and combination produce heat. Some of these elements, such as water and sodium chloride, do not undergo much alteration; others, again, are transformed into new substances which assist in reconstruction, while some favor destruction; the latter producing acidity in the tissues as well as the generation of heat, similar to that produced by muscular activity. Chemistry leads to the deduction that it is the action of oxygen on carbon which produces *oxidation* in the vital animal economy, similar to the action of the oxygen of the air in combining with the various fuels, such as wood, coal, oils, and illuminating gas, giving rise to what we call *combustion*, by which heat is generated. Any increased quantity of fuel and oxygen increases the heat of a furnace, or heat generator; but how can one satisfactorily explain, chemically, the rise of temperature as being oxidation in the living animal economy, in typhoid or any other fever, when no increased quantity of oxygen and carbonaceous food is furnished? All normal organic destructive tissue-changes produce acidity, which, coming in contact with the alkaline fluid (blood), changes the carbonic acid bases, the product stimulating the lungs, heart, and several other organs, such as the skin and kidneys, to continual *normal* activity, when the ready-formed gas issues through the lungs, skin, and kidneys; but when the molecular activity is increased, as in certain pathological conditions, then the carbonic acid products are increased, and in return stimulate all the cells of the tissues to *abnormally* increased activity, resulting in an elevated temperature, which may be lowered by opening all cutaneous vents (perspiration) and allowing the acid elements and gas to escape. This is the true physiological process in the lowering of temperature. All others, such as cold or sedatives, are artificial. In the artificial process the cells and organs become more or less paralyzed and their activity decreased. It may be said in opposition: It is not the carbonic acid elimination, but the elimination of heat, which reduces the temperature. If the latter were true, the temperature would rise again immediately when the elimination stops. This it does not do. To every action there is a reaction, so that when the cells in the tissues become anæmic, lamed (paralyzed) from overwork, the production of carbonic acid ceases correspondingly, similar to the manner in which excessive muscular and excessive mental action lessens its formation, while moderate exercise of either increases it. Here we may find the cause of the periodical rise of temperature, be it from alimentation, periodical formation or ripening of a poison within the system during a pathological condition, periodical activity of saprophytes or pathogenytes, or from other causes stimulating the cells to increased activity, increasing the evolution of carbonic acid, and, as a consequence, causing a periodical rise of temperature.

Now, in the foregoing there are sufficient grounds for asserting that the terms "oxidation," "combustion," or "a burning up," are ill applied when used in reference to the evolution of heat in the vital economy. Such terms tend only to confuse when referring to the *vital action* within the organism. The *metabolic* processes of cells and tissues (this includes *anabolic* and *katabolic* processes), taken as a whole, may resemble combustion, because oxygen and carbon are taken in and carbonic acid gas is given off; but (to illustrate) no one would call the mixing of soda and vegetable acid in water, even though carbonic acid gas be evolved, oxidation. Again, no one would call the mixing of equal parts of alcohol and water, although heat be generated, oxidation. In no instance in nature is the combustion of a substance containing three-fourths of its own quantity of water possible, whereas, the blood and most of the tissues do contain this amount. The terms "oxidation" and "combustion" applied to the vital economy are therefore improper.

Again, too much stress is laid on the importance of oxygen in the vital economy. It is true that without it vital processes could not go on; but the same thing applies to many other elements, such as carbon, hydrogen, and sulphur. An abnormal amount of pure oxygen introduced into the system will paralyze, by oversatisfying the respiratory organs, as completely as carbonic acid will paralyze them by overstimulation. Oxygen generally receives the credit of being the sole cause of the chemico-vital actions; but we find that if it were not for the carbonic acid in the animal economy, respiration and circulation would cease. We also find that the exhaled gas is not fully covered by the term "carbonic acid," or CO_2; consequently, the extent of action of the elements of said acid, while yet in the blood, must be of far-reaching importance in regard to stimulation, in fact more so than is generally believed; *i. c.*, of such as already above indicated.

Nitrogen, combined with hydrogen, oxygen, carbon, and sulphur forms important compounds (proteids) of the secreted solids and fluids within the animal body.

Physiologically, sulphur exists in all albuminous substances; and, as no animal organic life can exist without albumen, the question arises, Has sulphur a physiological importance? This question is introduced here simply with reference to oxygen, to show that other elements are fully as necessary as oxygen is to animal vitality. Sulphur is not an accidental addition, as it is constantly present and found in all animal tissues. These are built up principally from albuminous substances, all of which contain sulphur, no animal organism or tissue existing without it. It is also necessary for the elementary activity of cell-life, which it increases by assisting the chemical combinations of molecules of hydrogen, oxygen, nitrogen, and carbon; and, consequently, of albumen, which is thus dissolved, changed, and made capable of forming a part of the tissues. In comparison, just as the iron in hæmoglobin is necessary to enable the hæmoglobin to take up oxygen in the lungs, and then be carried to the tissues, so is sulphur primarily necessary for the continuation and transformation of albumen into tissues.

In the male, the production of carbonic acid increases from the age of seven to twenty-five, but beyond forty-two it gradually decreases. As old age approaches it may even not exceed the amount produced at the age of ten. In the female, it increases from the age of seven until menstruation, the production of carbonic acid then remaining stationary up to the cessation of the menses, when it decreases, as in the male. Males from fifteen to forty-two exhale nearly twice as much carbonic acid gas as females, and a child twice as much as an adult, if their disparity in weight be duly considered.

Certain animals that breathe by lungs, as ducks and frogs, are nevertheless able to remain under water for a considerable time, as their thin, moist, and flexible integument then performs the function of respiration. Their skin is not covered with dry cuticle, and while under water the lungs are inactive, but respiration goes on; i.e., the exhalation of carbonic acid gas and absorption of oxygen takes place through the skin.

Dyspnœa (difficult breathing) may be caused by lung disease preventing sufficient respiratory movements, partial closure of respiratory passages, insufficiency of red blood-corpuscles, affections of the respiratory muscles, nerves, or centre in the medulla oblongata, deficient circulation in the lungs from heart affections, especially mitral valve lesions, dropsy, uræmic poisoning, or from fever heat, producing excessive carbonic acid stimulation on the heart and lungs.

In order to promote respiration in the newly born infant, or to overcome asphyxia from drowning or from other causes, stimulation of the nerves which are distributed to the face or chest may cause powerful reflex action by the centres of the pneumogastric, spinal accessory, phrenic, and intercostal nerves. This fact is proven by the sudden application of cold water to the face or chest, or by alternate applications of hot and cold water. (*Vide* Artificial Respiraton in chapter on Miscellaneous.)

CHAPTER XI.

TEMPERATURE.

In the human body a standard temperature is maintained, notwithstanding atmospheric variations. Vital or animal heat is produced by the changes of chemical combinations of elements, including the activity of the nutritive processes and disintegration that take place separately in each organ and tissue; therefore, the temperature varies in each organ and tissue:—

1. The more active the organ the more blood passes through it; consequently, the higher the temperature the more active its tissue changes.

2. Protoplasm, in the process of its activity, evolves heat. The more active, especially when the process of disintegration surpasses that of construction, the higher the temperature.

3. The liver and other glandular organs are great heat-producing centres.

4. Mental and muscular activity increases heat.

5. The amount of heat abstracted from the surface of the body depends on the extent exposed and atmospheric influences.

Heat *is a sensible quality of bodies accompanied by motion.* It expands all kinds of matter, whether solid, liquid, or gaseous. Heat and energy are so related that a definite quantity of one may be always changed into a definite quantity of the other; for, according to the doctrine of the *conservation of energy*, when one form disappears another or others take its place, *i. e.*, the *correlation* of energy is such that one kind can be changed into another, and the sums total are equivalents. Molecules and their atoms possess *potential* energy, which becomes *kinetic* during chemical action; *i. e.*, the clashing together of atoms, in consequence of affinity, generates heat. Such is, in part, an explanation of the *origin* not only of *animal heat*, but also of *muscular motion* and *organic activity*. In animal organic activity, the vital principle also maintains its share. Heat is required for the normal molecular action of all vital activities, since heat causes the molecules to move more rapidly and to a farther extent in their spaces, usually pushing their neighbors away a very little, and the size of the body increases. Hence, heat will diffuse itself into surrounding tissues, and thus tends to equalize the temperature; always passing from a warmer to a colder section.

A liquid will dissolve a solid only when the adhesion between them is greater than the cohesion in the solid; and, as heat generally weakens cohesion more than it does adhesion, therefore an abnormally high temperature of the liquid blood dissolves the more solid tissues more rapidly and in greater quantity than when at a normal temperature. Heat is always generated in consequence of motion; so that from any irritation, molecular or mechanical activity, the temperature is raised, and the effect is expansion and change of matter. The temperature of the living body is the same in summer and winter, its regular standard being maintained by the blood-circulation and by radiation of heat, the equalizing or moderating power lying in respiration by lungs and cutaneous perspiration.

Absolute rest does not exist in nature, consequently not in the human body. Motion and rest, therefore, are only relative terms. By food man appropriates the latent energies of the animal and vegetable substances, which are converted into energy—atomic and molecular motion, heat, cell-growth, tissue changes, circulatory, respiratory, secretory, and muscular activity.

The deeper tissues of the body are warmer than the more superficial, due not only to the increased vital action and consequent changes taking place in those structures, but also to their greater protection, while the more external parts are constantly losing heat. Immediately after a full meal the heat of the body is reduced, but as soon as digestion commences and absorption takes place its temperature rises. It is diminished by abstinence from food and exposure to cold. The division of a sympathetic nerve causes a relaxation of the blood-vessels to which it is distributed, destroying the connection for reflex action, so that the blood pressure relaxes (dilates) the vessels, thereby increasing the quantity of blood in the part affected, which at first increases the temperature; but subsequent loss of nutrition of the part follows. The temperature of venous blood varies in proportion to the activity of each particular organ or tissue, while that of arterial blood is almost uniform, except in the arterial capillaries of the lungs and skin, where it is cooled by exposure to a cooler atmosphere, the body thereby constantly losing heat by radiation and conduction; therefore, this deficiency must be continuously overcome by newly formed heat in its interior. Wherever a combination of atoms of substances that have an attraction for each other takes place, heat is evolved, illustrated by mixing water three-fourths and sulphuric acid one-fourth, by which heat is instantly evolved. The chemical potential energy of the atoms is converted during collision and recombination into kinetic energy; *i. e.*, into molecular motion, so that any change of *vis viva* (living energy) in the system corresponds to an equal amount of exertion gained or lost by the attractions of the particles on each other. Through the medium of the blood the materials necessary for the renewal of the normal quantity of caloric are supplied—supplying heat also where lost by exposure. The temperature of the blood in the left auricle of the heart is from 1 to 6 per cent of a degree lower than in the right, on account of having been cooled by contact with the atmosphere in its passage through the lungs. In the hepatic veins it attains the highest temperature of the body on account of having passed twice through functional activities, such as the capillary circulation of the spleen, stomach, intestinal wall, and the liver; while in the blood of the peripheral capillaries it is the lowest. The average mean temperature of the venous blood is higher than that of any other tissue. When the internal temperature rises to from 107° to 110° Fahrenheit (41.5° to 43° Centigrade) it generally proves fatal, though the individual can live in an atmosphere of from 118° to 125° F. (47° to 50.5° C.). *Heat* is measured in calorics; *temperature*, in degrees. A *caloric* is the quantity of heat required to raise the temperature of 1 kilogram of water from 0° to 1° Centigrade. The term *temperature* has no reference to quantity of heat, for if a pint be taken from a gallon of boiling water, the temperature of the former is the same as the latter, although having only one-eighth the quantity.

The normal temperature in an adult is about 98.5° F. or 37.25° C.; but it is normally highest between 4 and 5 P. M. and lowest between 4 and 5 A. M. At birth it is about one or two degrees higher than between birth and 50 years, but at extreme old age it rises again.

Fig. 23.

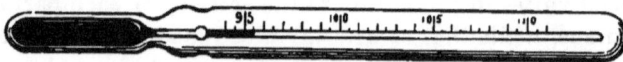

Diagram of Thermometer according to Fahrenheit.

To change Fahrenheit degrees into Centigrade, or vice versa:—From the given F. subtract 24, and half of the remainder will be the corresponding degree C.; or, to change Centigrade into Fahrenheit, double the given C. and add 24; the sum will be the degree F.

The normal temperature of the body may be maintained from two sources when exposed to cold; viz., ample supply of food, and muscular activity. The ill fed cannot resist cold. A persistent abnormality indicates some morbid action in the economy, though a fatal disease may exist even with a normal temperature, as is often the case in chronic Bright's disease.

Temperature is intimately related to the functions of vitality. If it be increased or diminished by a few degrees, a corresponding interference with the perfect mechanism of the various organs occurs, so that their relative activity may either be augmented or arrested. The phenomena of vital activities, such as nutrition, growth, and reproduction, are possible only within certain limits of temperature; for as soon as it approaches the limit below or above the normal, constructive vitality ceases. These limits, or extremes, of internal human temperature range between 90° and 110° F.—i. e., about 20 degrees; but it is difficult to give a general extreme in any special disease, as the fatal point depends on the pathological condition of the system at the time. If an abnormally high temperature is not checked, either the coagulation of the albuminoids and amyloids of the protoplasm, or of the proteids of the tissues, or of both, takes place; there is then an increased change or decomposition of the tissues and secreted fluids, which results in an arrest of nutrition and vitality, and causes death.

Generally, temperature rises or falls in unison with the rise and fall of the pulse. For every degree above 100° F. the pulse rises ten pulsations per minute. It may be several degrees above or below the normal for some days before sickness is experienced; consequently, the importance of an early investigation of internal temperature is apparent, as it often furnishes the clue to correct diagnosis of a disease which cannot otherwise be surmized, and is an early guide for treatment. The different kinds of food and stimulants have their respective influences on the temperature; consequently, it is important to make more than one observation, and at different times. It is also a sure detective in certain cases of feigned disease; or if an increased pulse is due to fever or to anæmia; or if abnormal sweating is the result of external heat, activity, disease, or exhaustion. If from disease, it will be preceded by elevated temperature.

In regard to internal temperature the external indications cannot be relied upon, as the integument may feel cool to the touch though the thermometer may register 105° or more, as in some low forms of disease; consequently, external indications simply are not a means by which we can acquire any definite idea of the temperature of the body. Illustration: if a piece of flannel and a piece of iron are left in a room for some time, each will be of the temperature of the room; yet to the touch the flannel would be pronounced warm and the iron cold. The temperature of the body may be taken by placing the thermometer under the arm-pit or the tongue, but it is most accurately indicated by placing it for 8 or 10 minutes in the rectum or vagina. No cause except disease or old age changes the temperature more than 1.5° F. from the normal. The excessive waste of tissues in old age often causes the temperature of the body to rise two degrees above normal without exhibiting any signs of disease, and these cases would have to be considered normal. A persistent increase of temperature of the body without exhibiting any particular local or general disease indicates tuberculosis somewhere in the system. An increase during tuberculosis indicates that the disease is advancing, with new complications forming. In pneumonia a temperature which never goes beyond or higher than 102° F. does not indicate soft infiltration in the lungs; but a persistent temperature of 104° or more indicates a severe attack. In acute rheumatism 104° is an alarming symptom. In exanthematous diseases, such as measles or scarlatina, the fever rises till the breaking out or eruption; the same takes place in small-pox till the pustules appear and burst. In tetanus, when all the muscles are in spasms, the very high temperature probably depends on the increased decomposing changes of tissues produced by the excessive muscular activity. The decomposed tissues probably produce ptomaines in which pathogenytes develop, constituting the tetanus bacilli found. In jaundice an increase of temperature above the normal indicates a pernicious turn. In kidney disease a normal temperature indicates chronic Bright's disease.

When the internal temperature is below normal, accompanied by an abnormally low pulse, there is possibly either some pure bile in the blood paralyzing the sensitiveness of the endothelial cells of the heart, diminishing the afferent (sensory) impulse for reflex functions to the vaso-motor and inhibitory nervous centres; or pressure on these centres, cardiac plexus, or nerves, by a new growth; or a cancerous condition somewhere in the system, most likely in the liver, or in a neighboring part affecting the liver. A case is reported in which the temperature of a person in apparent health suddenly fell to 93° F. from intestinal hæmorrhage, though the blood did not appear in the stools for several days after.

In fevers the temperature varies from 101° to 106° F. If it exceeds the latter the patient is in danger, unless due to a malarial affection, in which case it may rise to 105° quickly, although the patient may have been apparently well shortly before. Such rise usually commences several hours before the beginning of the ague paroxysm. If a person recently healthy exhibits a sudden rise of temperature to 104° or 105° F., it is usually only an attack of ephemeral, exanthematous, or incipient malarial fever, but not typhoid. In the latter the rise of temperature is gradual, being about 1° daily above the preceding day until it reaches 105°. The fever is at its highest daily at 4 P. M. and 1 A. M., and lowest at about 9 P.M. and 5 A. M. If higher in the forenoon than the previous afternoon it is an indication that the disease is more severe. At any period in the course of typhoid a temperature of 1C6° is alarming, a persistent rise to 107° generally terminating in death. A sudden rise to from 105° to 107°, or sudden fall below normal temperature, usually indicates collapse; this may happen some days before any change in the pulse or other signs are observable.

The most direct and simple means of moderating high temperature is by increasing the activity and secretion of the perspiratory glands; *i. e.*, by *cutaneous perspiration.* The quantity of watery vapor discharged daily from the whole body is about 43 ounces. Of this amount the skin gives off about 26 and the lungs 17 ounces; so that the heat given off normally by evaporation is about one-fifth the total produced in the body.

In growing plants heat is actively produced, but very rapidly dissipated owing to the extent of surface exposed to evaporation by their leaves and branches. If this loss be artificially limited, the elevation of their temperature becomes apparent, and may be estimated. During germination and blooming the development of heat is much more rapid. (For additional information on temperature, see chapters on Respiration, Circulation, and Medulla Oblongata.)

CHAPTER XII.

THE NERVOUS SYSTEM.

In structure, the nervous system consists of gray and white substances and nerve-fibres. The *gray substance* is composed of collections of nervous centre cells intermixed, or mingled, with nerve-fibrillæ. It constitutes the external portion or convolutions and various ganglia in the central part and about the base of the brain, the central portion of the spinal cord, and isolated ganglionic masses in the sympathetic system. These latter ganglia are found in different parts of the body. Every collection of gray substance is termed a *nervous centre*. The *white substance* of the brain and spinal cord is composed of nerve-fibrillæ, and is found in the internal portion of the brain and external portion of the spinal cord.

Anatomically and physiologically, the nervous system comprises two great divisions—the Cerebro-spinal and the Sympathetic.

The **Cerebro-Spinal System** consists of the brain, spinal cord, and nerves emanating therefrom. It possesses the function of bringing not only its own organs, but also all others of the body, into relation one with another, so that all parts are co-ordinated in one harmonious whole, while some of its organs are concerned in mental processes in a manner to be presently explained. On its organs, sensation and motion also partly depend. Its nerve-fibres are distributed to the organs of the senses, to the voluntary muscles and the integument, as well as to the external openings and terminations of internal passages. They also furnish the principal nerve supply to the heart, muscles of respiration, and some other involuntary muscles.

The involuntary muscles and the various layers of muscular fibres in the internal organs are not directly dependent on the cerebro-spinal system, as they owe their power of contraction to the influence of the sympathetic system and possess this power to some degree within themselves. (For the action of the sympathetic system, see the chapter on Sympathetic System.) The bladder, stomach, and intestines, for example, are enabled to act upon their contents, partly through their own inherent function of muscular movement, and their direct communication with the sympathetic system.

Some muscles are both voluntary and involuntary in their function, and are termed mixed, as, for instance, those of respiration.

The function of the nervous system is to associate the different parts of the body into one harmonious whole and to bring the various organs into mutual relationship. This correlation of function is termed *sympathy*, and is accomplished by either a physical or mental impression from a direct or indirect stimulus, and may affect only a part of or the entire body—physically, as by an electric current; or as a solution of atropia upon the cornea produces dilatation of the pupil by paralyzing the muscular fibres of the iris. A physiological action becomes manifest when food is taken into the stomach, causing the secretion of gastric juice. The digestion of food stimulates the gall-bladder to empty itself into the duodenum; the presence of food in the mouth, the act of mastication, or even the aroma of a favorite dish, each effects the secretion of saliva by the parotid, submaxillary, and buccal glands, together known as the salivary glands. Mental excitation may produce perspiration.

The **Brain**, or **Encephalon**, comprises all that portion of the

Fig. 24.

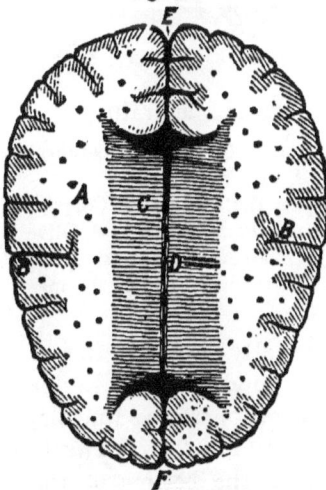

Diagram of Corpus Callosum.

A, White substance of the cerebral hemispheres. B, Fissure of Sylvius. C, Corpus Callosum. D, Longitudinal fissure, extending from F to E. E, Posterior aspect of the longitudinal fissure. F, Anterior aspect of the longitudinal fissure.

nervous substance within the cavity of the cranium. It is divided into the cerebrum, cerebellum, pons Varolii, and medulla oblongata. These are subdivided, forming a symmetrical series of minor nervous centres, connected with each other and the spinal cord by tracts of white substance; *i. e.*, by nerve-fibrillæ. The longitudinal fissure (D, E, F, Fig. 24) divides the brain into two equal hemispheres.

The principal brain centres (Fig. 25) are connected by intermediate nerve-fibrillæ, though each centre has special functions independent of the others. Again, the cerebral areas (centres) have special connections with sympathetic ganglia; as, for example, a person may be blind with one or both eyes, with or without loss of normal movements of the iris, eyes, or other functions. This is true of all sense organs. The greatest quantity of gray substance is at the sulci, or grooves (fissures), of the brain convolutions. The convolutions, as well as the fissures, differ in each individual brain; but four fissures are similar in all human individuals: viz., the longitudinal, Sylvian, Rolando, and parietal.

Fig. 25.

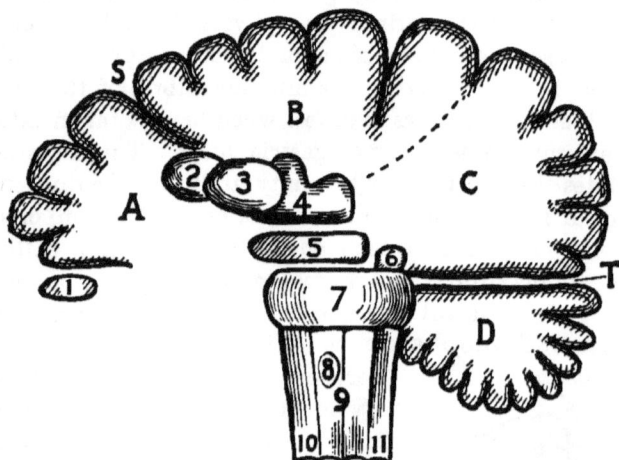

Diagram of the principal nervous centres of the brain.

A, Anterior cerebral lobe. B, Middle cerebral lobe. C, Posterior cerebral lobe. D, Cerebellum. T, Tentorium cerebelli. S, Sylvian fissure.

1, Olfactory bulb.   2, Corpus striatum.   3, Optic thalamus.   4, Corpora quadrigemina. 5, Crus cerebri. 6, Pineal gland. 7, Pons Varolii. 8, Olivary bodies of the medulla oblongata. 9, 10, and 11 ,Medulla oblongata. 10, Anterior pyramid of the medulla oblongata. 11, Restiform body of the medulla oblongata.

Each cerebral hemisphere is divided, especially on its inferior or under surface, into three lobes—the anterior, middle, and posterior (A, B, C, Fig. 25).   In pressing apart the hemispheres we find the *corpus callosum*, or *commissure* of the brain (C, Fig. 24).   This is a thick, white stratum of transverse fibres at the bottom of the longitudinal fissure, connecting the two hemispheres of the brain. It is about four inches long, forming the roof of the lateral ventricles.   These ventricles have prolongations, or horns, extending into the three lobes.   Beneath, and almost inclosed by the corpus callosum, is the fifth ventricle; below this is the third ventricle (3, Fig. 26), which communicates downwards by a narrow passage with the fourth ventricle in the medulla oblongata.

The chemical composition of a healthy adult brain is given in the following table:—

Water	about	73	per cent.
Fat	"	12	" "
Albuminous matter	"	8	" "
Salts	"	5	" "
Phosphoric acid	"	2	" "
Total		100	

In insane persons the salts have been found to be only about 2½ per cent.

The supply of blood to the brain is by way of four arteries—the two internal carotid and the two vertebral arteries. Their intercommunication, through the *circle of Willis* (Fig. 27), enables the brain to receive its proper amount of nutrition, even should one artery only convey the blood to the brain. If the circulation is interfered with or stopped for a moment, insensibility immediately follows, whether it be from disease of the cerebral arteries or from embolism.

Hæmorrhage upon the surface of the cerebral hemispheres produces coma, but if hæmorrhage occurs in the substance of the cerebral ganglia — the corpora striata or optic thalami—it produces hemiplegia.

Galvanization of the dura mater on one side of the head produces muscular twitching on the same side of the body; but galvanization of the cerebral convolutions on one side excites muscular action on the opposite side of the body. This is because of the decussation of the nerve-fibres at the upper part of the spinal cord. As a rule, neither the white nor the the gray substance of the hemispheres is excitable under ordinary artificial stimulus, unless it reaches the fibrous bundles of the deeper parts of a ganglion.

Fig. 26.

Diagram of the ventricles of the brain.

1 and 2, Lateral ventricles. 3, Third ventricle. 4, Fourth ventricle. 5, Fifth ventricle. 6, Sixth ventricle, or central canal of the spinal cord. 7, Foramen of Monro. 8, Aqueduct of Sylvius.

The **Cerebrum** is the largest portion of the encephalon, or brain, forming the whole upper and anterior part (A, B, C, Fig. 25). The longitudinal fissure divides it into hemispheres (E, F, Fig. 24). The upper surface of the cerebrum is thrown into winding convolutions, separated by deep grooves (sulci), into which processes from the pia mater project. The number and extent of the convolutions bear a relationship to the intellectual power of the individual. Some of the lower animals have many cerebral convolutions, but they are shallow; while others have few or none; the brute follows his instincts, but asks not the reason of them. The exterior of the cerebrum is a gray, cortical, bark-like substance; the interior consists of ganglia of gray, and white medullary substances. The nervous centres belonging to the cerebrum are as follows: the gray convolutions covering the anterior, middle, and posterior lobes (A, B, C, Fig. 25); the olfactory lobe or bulby enlargement of the olfactory nerve (1) being concealed beneath the anterior portion of the anterior cerebral lobe; corpus striatum (2); optic thalamus (3); corpora quadrigemina (4); crus cerebri (5); and pineal gland (6).

Fig. 27.

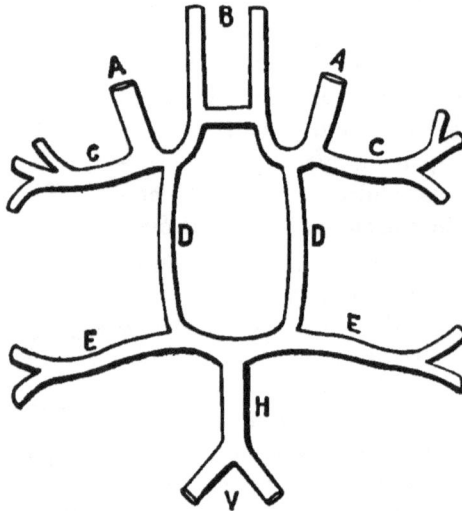

Diagram of the Circle of Willis.

A A, Internal carotid arteries. B, Anterior cerebral arteries. CC, Sylvian arteries. DD, Posterior communicating arteries. EE, Posterior cerebral arteries. H, Basilar artery. V, Vertebral arteries.

On the cells of centres of the cerebral convolutions, the operations of the intellect in man's present state partly depend. This includes internal and external sensations, inasmuch as to each external sense-organ a proper centre-cell in the cerebrum is united, without which there is no external sensation; internal sensation, inasmuch as each external sense must have its own internal organ, as will be shown afterwards; the operations of the intellect, inasmuch as the intellect, although a faculty of the soul, and therefore inorganic, cannot act in its present state—*i. e.*, while the soul is united to the body—unless the operation of the imagination (one of the internal senses, having its organ in the cerebrum), depending primarily on the external senses, precedes and accompanies it.

Physiologically, the cerebrum is not directly essential to life; since respiration and circulation are but indirectly governed by the cerebral centres. Life may, therefore, continue for a time even if the cerebrum be diseased, injured, or partially destroyed, as long as the respiratory and circulatory centres of the medulla oblongata remain unaffected. A lesion of the cerebral convolutions affecting the gray substance, or an impairment of its functions, causes either a partial or total suspension of the mental functions, for the reason above assigned: such as *reason*, which is an act of the mind by which is determined the agreement or disagreement, the identity or diversity, of two things, by comparing them with a third; and *judgment*, an act of the mind by which it affirms the agreement or disagreement of two concepts or ideas.* When these functions, depending on cerebral activity, are abnormal, then we say that the individual so affected is *insane* and incapable of performing legal requirements and duties; and his responsibility to *God* and *man* ceases. *Insanity,* therefore, may be properly defined: *Impairment of cerebral organs affecting reason and judgment.* Insanity may exhibit itself in *monomania, mania,* or *dementia.* As previously stated, it has been claimed that the salts of the brain are present in a deficient quantity; though abnormal functions of cerebral areas may manifest themselves even without visible physical lesion being observable after death.

---

* "It is one and the same power of the soul that understands, reasons, and judges. That power which is concerned in the acts of *knowing*, be it manifested as understanding, reason, or judgment, is what is generally called the *mind.* It is, however, sometimes considered as the three powers, or faculties, of the soul taken together—the *will, memory,* and *understanding*; the *soul* being the spiritual substance, together with its perfections."

9

The **Corpora Striata**, (2, Fig. 25; and C, Fig. 28), one in each cerebral hemisphere, are partly separated by the two optic thalami. Their gray masses are striated with white fibres, hence their name. Each corpus consists of gray matter imbedded in the white substance of said hemisphere. In the center the corpus contains a layer of white matter (nerve-fibrillæ) dividing its gray matter into two portions; the internal portion is termed *nucleus caudatus*, the external portion *nucleus lenticularis*. A layer of white matter separates the latter from a body of gray mater termed *claustrum* and from the *island of Reil*. The function of the claustrum is not known.

Fig. 28.

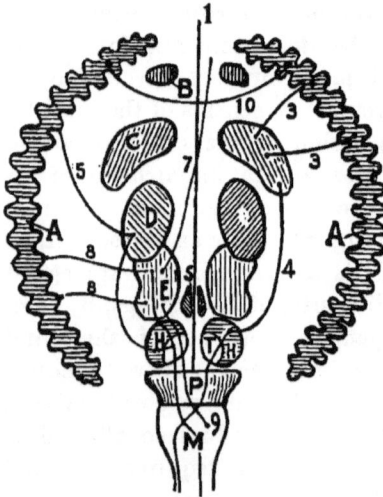

Diagram of the brain centres.

AA, Cortex and cerebral convolutions. B, Olfactory bulb. C, Corpus striatum. D, Optic thalamus. E, Corpora quadri-gemina. S, Pineal gland. H, Crura cerebri. T, Tegmentum and H, Crusta of the crura cerebri. P, Pons Varolii. M, Medulla oblongata.

1, Longitudinal fissure. 3, Place of the corona radiata and lenticular nucleus— the lines indicate motor fibrillar communication. 4, Motor fibrilla tract. 5, Sensory fibrilla tract. 7, One optic nerve fibrilla. 8, Fibrillar communication of sensory area of sight. 9, One oculomoto-rius fibrilla. 10, One commissure fibrilla.

Motor fibres from the convolutions (3, 3, Fig. 28) proceed to the corpus, then to the anterior columns of the spinal cord; since voluntary impulses emanating from the convolutions pass to the cells of this corpus, where they may be modified, reinforced, or brought into harmoniously acting motor impulses (C, Fig. 28). This centre is in communication with voluntary muscles; and if the corpus striatum on one side of the brain be injured, hemiplegia of the opposite side of the body results, on account of the decussation of the nerve-fibres. Impairment of both corpora may lead to permanent loss of voluntary motion of the body.

The **Optic Thalami** (*thalamus*—a bed), one in each cerebral hemisphere (2, Fig. 25; and D, Fig. 28), are situated below the corpus callosum and partly between the corpora striata, separating these bodies. They are two oblong masses of gray substance, having in the center some white matter consisting of nerve-fibrillæ, forming the communication with the more centrally located cells of the gray substance. Sensory fibres of the posterior columns of the spinal cord proceed to the optic thalami (D, Fig. 28), thence to cerebral convolutions. The thalamus is the modifying and distributing centre between the brain areas of sensation and peripheral sensory organs. Consequently, if in one hemisphere it is destroyed, *hemianæsthesia* of the other side of the body results, on account of the decussation of the nerve-fibres. The corpora striata and optic thalamus together are sometimes termed the *basal ganglia*. The conflicting results of experiments in regard to local stimulation, or lesion of either one of these two ganglia, in reference to loss of sensation, motion, or both, may possibly be accounted for by the close relationship of the ganglia and their nerve-fibrillæ.

The **Corpora Quadrigemina** consist of two rounded tubercular eminences in each cerebral hemisphere, and are separated by a slight depression (4, Fig. 25; and E, Fig. 28). They are situated behind the third ventricle and beneath the posterior portion of the corpus callosum, laterally joining with the optic thalamus and the upper surface of the crura cerebri. They are solid bodies, composed of white matter externally and gray within. In function they govern the movements of the iris as well as the harmonious co-ordination of the movements of both eyes. If the function of one of these bodies becomes impaired, the iris on the opposite side cannot contract; or, if totally destroyed, sight is lost, although, apparently to the observer, the eyeball has not undergone any pathological change. Its irritation simply produces contraction of the iris. In impairment of the corpora quadrigemina the equilibrium of the body is interfered with. (See Semi-circular Canals.)

The **Pineal Gland**, so called from its pine-cone shape, lies on the median line beneath the posterior portion of the corpus callosum, partly above and behind the third ventricle (6, Fig. 25; and S, Fig. 28). It projects backward and downward between the tubercula quadrigemina. Its function is not known, although the gray matter connected with it and projecting to the medulla oblongata may possibly be an auditory ganglion, associating it with the organ of hearing and the auditory area in the cerebrum, in the accomplishment of the sensation of sound.

The **Crura Cerebri** are two thick strands, or bundles, of white substance (5, Fig. 25; and H, Fig. 28); each crus being about three-quarters of an inch in length and resembling a leg (hence its name). In the center of each crus is a mass of dark-gray matter termed the *locus niger*, from which some fibres of the oculomotorius, or third nerve, originate, serving for numerous and complicated movements of the eyeball. The posterior portion of the crus is termed the *tegmentum* (T, Fig. 28), which is the path for the sensory fibres. The anterior part is called crusta (H), and is the channel for the motor fibres; consequently, it is at the crura cerebri where the spinal sensory and motor fibres enter and depart from the cerebrum. Each class of fibres forms a separate tract; the sensory, from the posterior part of the spinal cord, proceeding from the crura to the optic thalami (D, Fig. 28); thence (5) to the cerebral convolutions. The voluntary motor fibres (3, 3,), from the cerebral convolutions, proceed to the corpora striata (C); thence to the crus cerebri and cerebellum; thence, some to the voluntary muscles of the head, others to the anterior portion of the spinal cord and the organs of the body and limbs.

The **Valve of Vieussens** is a medullary layer in the Sylvian aqueduct, or canal, leading upward and forward from the fourth to the third ventricle (8, Fig. 26), and forming a part of the roof of the fourth. At the side of the valve originates the Trochlearis (Patheticus) nerve (fourth).

Below the cerebrum follows the pons Varolii (7, Fig. 25; and P, Fig. 28); while behind the pons and posteriorly below the cerebrum is the cerebellum (D, Fig. 25); anterior to the latter and below the pons, the medulla oblongata.

The **Pons Varolii** (*pons*, a bridge), is sometimes called *tuber annulare*. It is a thick, flattened band of alternate layers of transverse and longitudinal white fibres, with an intermixture of gray matter, and connecting as by a bridge the cerebrum, cerebellum, and the medulla oblongata (7, Fig. 25; and P, Fig. 28.) It is divided in the median line by the longitudinal fissure, each half containing centre cells which give origin of some fibres of the fifth, sixth, and eighth cranial nerves. Consequently, it is a nervous centre, and at the same time affords a passage for tracts of nerve-fibrillæ to and from the cerebrum, cerebellum, and medulla oblongata.

The **Cerebellum** or small brain (D, Fig. 25; and O, Fig. 29), is beneath the posterior lobes of the cerebrum, and separated from it by a membrane called the *tentorium cerebelli* (T, Fig. 25). Like the cerebrum, it is convoluted and divided into halves by the longitudinal fissure. The cortex of its convolutions consists of gray matter, while its interior is white. The convolutions are smaller than those of the cerebrum. In each cerebellar hemisphere the white matter contains in its center a dentated nucleus or lump of gray matter, the *corpus dentatum*, in the center of which is inclosed some white matter of nerve-fibrillæ, thereby communicating with the surrounding white substances. Some of the fibres of the cerebellum communicate with the lateral columns of the spinal cord, enabling the head to maintain the proper posture on the trunk. The cerebellum is a part of the motor apparatus, and, as it is the reflex centre which governs co-ordinate action, a lesion of its gray substance produces not only inharmoniousness of voluntary movements of muscles and limbs, but also disturbs the equilibrium in balancing of the body. It is not the centre which governs sexualism, as was formerly supposed. Any slight injury to the cerebellum may cause constant agitation of the head, limbs, or of both. (See Semi-circular Canals.)

CHAPTER XIII.

MEDULLA OBLONGATA.

The **Medulla Oblongata** is the superior enlarged portion of the spinal cord, situated within the cranium, extending from the upper border of the atlas to the lower border of the pons Varolii, and forming a part of the encephalon (8, 9, 10, 11, Fig. 25; M, Fig. 28; and M, K, L, R, Fig. 29). It is divided in the median line longitudinally, each half being subdivided by grooves into four columns — the anterior pyramid, lateral tract, the restiform body, the posterior pyramid. The anterior pyramid is a continuation of the anterior column of the spinal cord. The lateral tract (embracing the olivary body) is a continuation of the lateral column of the cord. The restiform body is a continuation from the posterior column of the cord inferiorly, and extending to the cerebellum superiorly. The posterior pyramid is a thin, cord-like continuation of the posterior median column of the cord.

The **Olivary Body** (8, Fig. 25; and L, Fig. 29), so named from its olive-shape, is situated on the lateral tract, behind the upper portion of the anterior pyramid. It contains masses of centre-cells. In the groove between the olivary and the restiform body arise some fibres of the glosso-pharyngeal, and at the groove between the olivary and pyramidal bodies arise some fibres of the hypoglossal nerve.

The **Fourth Ventricle** is a lozenge-shaped space in the medulla oblongata, formed by the expansion of the central canal of the medulla and the convergence of the posterior pyramids (4, Fig. 26). The floor of the ventricle is formed by the posterior surface of the medulla, the roof by the *valve of Vieussens* at the aqueduct of Sylvius (8) and anterior inferior surface of the cerebellum. This ventricle communicates upward by said aqueduct with the third ventricle.

In the lower part of the anterior and internal aspect of the medulla most of the nerve-fibres of the two hemispheres of the brain cross obliquely and in bundles to the opposite side of the spinal cord. This crossing is termed "decussation of the pyramids."

The medulla oblongata is a continuation of the spinal cord, although placed just within the cranium (evidently for protection from injuries to which the cord is more liable). It is the most important portion of the brain for the immediate continuation of life, as it contains the reflex centres presiding over respiration and circulation. It is that portion of the brain giving origin to nerve-fibres, more or less, of all cranial nerves, except the olfactory and the optic, and these even are indirectly connected with it. Consequently, in the centres of the gray matter of the medulla oblongata, the reflexes are effected for respiration, circulation, digestion, absorption, assimilation, secretion, and intestinal and capillary peristalsis. It also supplies nerve-fibres to the muscles and skin of the head and face, all mucous membranes, secreting glands of the skin, and other secreting organs. It likewise affords some fibres for three organs of the external senses; viz., *taste*, *sight*, and *hearing*.

Fig. 29.

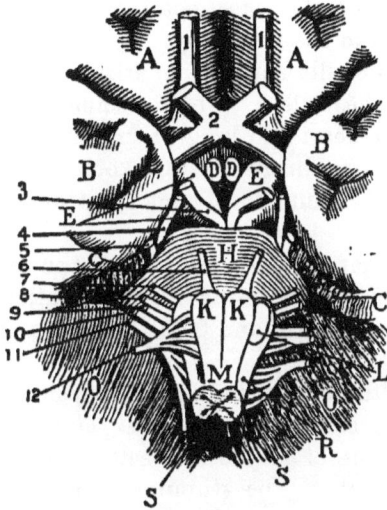

Diagram of the medulla oblongata and cranial nerves.

A, Anterior cerebral lobe. B, Middle cerebral lobe. C, Part of posterior cerebral lobe. D, Mammillary eminences. E, Crura cerebri. H, Pons Varolii. K, Pyramidal bodies. L, Olivary body. M, K, R, L, Medulla oblongata. O, Cerebellum. R, Restiform body. S, Section of the medulla from the spinal cord. 1, Olfactory nerve. 2, Optic commissure, formed by the optic tracts. From this decussation outward the bundles of fibres are termed the optic nerves. 3, Oculomotorius nerve. 4, Trochlear nerve. 5, Trifacial nerve. 6, Abducens nerve. 7, Facial nerve. 8, Auditory nerve. 9, Glosso-pharyngeal nerve. 10, Pneumogastric nerve. 11, Spinal accessory nerve. 12, Hypoglossal nerve.

The medulla oblongata is really the central point from which impulses are reflected affecting the whole vital organism; it contains the vaso-motor centre which influences the whole vascular system by controlling the dilatation and contraction of the blood vessels; the centre from which the motor impulses are reflected to enable the muscles of the palate, pharynx, and œsophagus to produce successive co-ordinate and necessary movements for deglutition, or swallowing; the centre by which the various muscular movements made in speech are co-ordinated; the salivary excitor centre; the cardio-inhibitory centre for the regulation of the action of the heart. Kirkes mentions "the so-called diabetic centre, which, if irritated in the medulla oblongata, causes glycosuria, this peculiar result being due to the vaso-motor changes and increased flow of blood to the liver." The diabetic result may possibly be brought about from the irritation of the vaso-motor centre-cells giving rise to nerve-fibrillæ passing to the liver only.

The blood flowing through the medulla supplies nourishment in order to keep up the energy of the cells. A high temperature acts as a powerful stimulus on the heart and lungs, increasing the stimuli for reflex action of the centres in the medulla oblongata. Cold lowers the stimulation, the reflex functions, and the temperature.

Normally, the vaso-motor and respiratory centres receive impulses for reflex activity — the former from the beginning of the formation of the *fœtus in utero*, the latter from birth, and both to the end of life.

The **Respiratory Centre** is situated at the *Calamus Scriptorius*, at the floor of the fourth ventricle. It is termed the *vital point*, and consists of two slight protuberances of gray centre-cells in each hemisphere of the medulla oblongata. The stimulation to this vital point is derived from the lungs, which are stimulated by the carbonic acid of the venous blood in the lungs. The oxygen of the inhaled air produces a satisfying influence. Consequently, for normal respiratory actions it is required that carbonic acid and oxygen be present in moderation. If the lungs are oversaturated with the elements of either one (of carbonic acid or of oxygen) the stimulation to the lungs and to the respiratory reflex centre ceases, and *apnœa*, or cessation of the respiratory movements, results. Respiratory stimulation, producing over-excitation of the respiratory centre, gradually exhausts it, resulting in *asphyxia*. This may happen even to the embryo in the uterus, if the mother's blood becomes deranged in that manner. Respiratory irregularity is one of the most minacious symptoms in brain affections, showing that the reflex centre of the medulla oblongata is affected. This phenomena depends on an improper reflex action of the medulla to the respiratory organs. A suspension of either sensation, volition, or intellectual power does not necessarily indicate immediate danger of death. Cerebral apoplexy, either from hæmorrhage at the surface of the hemispheres or a hæmorrhage in the cerebral ganglia, is not at once fatal; but it is so when occurring at the floor of the fourth ventricle in the medulla.

Deglutition is under the reflex control of the medulla, but mastication is voluntary, continued or arrested at will, while swallowing or deglutition is voluntary only from the mouth into the pharynx, after which it is reflex and involuntary; and, when commenced, cannot be arrested at will, and may therefore be performed after all voluntary power has disappeared. Swallowing, when no longer possible, or accompanied by choking during disease, and this when the throat is clear, indicates that the reflex centre of the medulla has become seriously affected, and death near at hand. Exercise of the voice or exhibition of emotion requires the co-operation of the cerebrum and pons Varolii; but the immediate mechanism in the larynx by which sound is produced has its nervous centre in the medulla. The fibres of the *hypoglossal nerve* originate in the medulla, and carry the motor impulses to the muscles of the tongue, and regulate the condition of the rima glottis. Disease of said nerve paralyzes these muscles, and makes articulation difficult or impossible. In this affection, the inability to speak is very evident from *ataxic aphasia*, which is of cerebral origin, consisting in loss of power in the necessary nervous combinations.

CHAPTER XIV.

NERVOUS CENTRES AND REFLEX ACTION.

In the nervous system every collection of gray matter is termed a **Nervous Centre.** A nervous centre is an area or aggregation of gray cells, with functions peculiar to each group; so that a mass of said matter generally has many centres, with various functions. In fact, every nervous centre concerned in reflex impulses has the power to inhibit, restrain, or accelerate them. The cells of nervous centres are round or irregular in shape, and consist of a soft, semi-transparent, finely granular albuminous substance, their size varying from 8 to 130 micromillimetres in diameter. The largest centre cells are found in the anterior horns of the spinal cord, the smallest in the sympathetic ganglia, the middle-sized in the brain centres. Most centre-cells of the cerebro-spinal system have processes communicating with the axis-cylinders of the nerve-fibres. Some cells have a single process, others more, and are severally termed *unipolar, bipolar, and multipolar* (Fig. 30). The nucleus of the cell is distinct, and contains a well-defined nucleolus in direct communication with the axis-cylinders of minute nerve-fibrillæ (2), which communicate either with another nervous centre-cell, or in a direct manner with peripheral termination cells, such as a tactile corpuscle, Pacinian body, terminal bulb, or a terminal plate. (See chapter on Nerve-fibres and Their Terminals.) Some of the nervous centre-cells have a process which terminates abruptly (4), appearing to have no communication with nerve-fibrillæ. Such a process is termed an *axis-cylinder process*, the function of which is unknown.

Fig. 30.

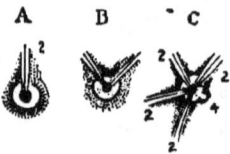

A   B   C

Diagram of nervous centre-cells.

A, Unipolar cell. B, Bipolar cell. C, Multipolar cell. 2, Process through which the nerve-fibrillæ pass to or from the cell. The axis-cylinders of the nerve-fibrillæ are in communication with the nucleolus of the cell. 4, Axis-cylinder process.

The manifestations, or functions, in which the different nervous centres are involved, either singly or in combination, may be said to constitute nine classes—five for the external senses, and four for the internal. The centres involved with the five external manifestations have all been determined to be connected with the organs of *sight* (eye), *hearing* (ear), *smell* (nose), *taste* (tongue and mouth), and *touch*. The centres of the four internal sense-organs, the locations of which have not yet been determined, are those for the manifestation of *common sense*, the *imagination*, *estimative faculty* or instinct, and *sensitive memory*. (See chapter on Senses.)

The impulses which call the nervous centre-cells into action are derived from impressions by stimuli on either the external or internal organs, through sensory nerve-fibres. The nervous centre, then, directs, distributes, or forwards an impulse to either peripheral or internal organs direct, or to another nervous centre; the latter to either direct, modify, analyze, classify, redistribute, or check them, and to convert them into impulses which are transmitted through motor fibres to distant organs. Some groups of centre-cells act constantly or periodically. Of those acting constantly the vasomotor and respiratory centres in the medulla oblongata are good examples. It is well known that with volition in suspense, as when asleep, actions by reflex are much more severe; in fact, when awake, many can be entirely prevented. Slight epileptic attacks, and convulsions of children from indigestion, may often be prevented by will power during waking hours, which during sleep result in severe attacks. Stimulation of a sensory organ may produce various reflex actions simultaneously, especially in children. For instance, powerful stimulation of the skin of the foot may cause movement, in bending and twisting the limbs and body; secretion, such as tears; excretion, such as urine or sweat; or pain in the organ stimulated. Powerful stimulants increase the reflex excitability, while narcotics diminish it.

Reflex, physiologically considered, takes place only in a nervous centre, and refers to the action of an organ (muscle, respiratory, circulatory, or secretory organ) receiving its primary impression from stimulus directly, or indirectly through impulses from another organ. The terminal sense-organs do not transform the outer energy into physiological energy, but the stimulus excites changes in the terminal organs in producing activity within them. The individual may or may not be conscious of this action. *Sensation* is the vital organic representation. (See chapter on Senses.) *Property* applies to the endowment of a nerve, nerve-fibre terminal, or muscular fibre, to respond to a stimulus; the action of response is their *irritability.* Irritation is always followed by the phenomena noticeable in living matter only; but does not involve the exercise of the inherent force of the irritant itself. *Excitability* is the special irritability of living matter, and may be affected by a variety of stimuli: —

1. Normally, by influences on, or changes in, the terminal or central organs.

2. Thermally, by variation of temperature.

3. Chemically, by the application of acids, alkalies, metallic salts, etc.

4. Electrically, by continuous or induced currents.

5. Mechanically, by intermittent pressure, pricking, beating, section, etc.

Heat increases, while cold diminishes, excitability. When a nerve or a muscle is at absolute rest for a long period its excitability diminishes; or, if such tissue be inactive beyond a certain time, it wastes, becoming thinner, and degeneration of its own substance results.

Fig. 31.

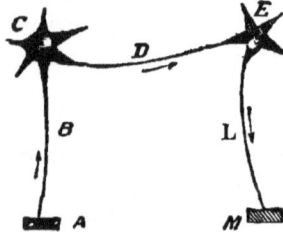

Diagram of the reflex act.
A, Sensory receiving organ. B, Afferent (sensory) nerve-fibre. C, Nervous centre-cell. D, Nerve-fibrilla between centre-cells. E, Nervous centre-cells. L, Efferent (motor) nerve-fibre. M, Muscular, secreting, or excreting organ.

Reflex action requiring the co-operation of cells of several centres in order to act harmoniously towards certain results is termed *co-ordination.* Either voluntary or involuntary external activity is, primarily, involved in reflex action, which essentially requires seven agents: a sensory receiving organ (A, Fig. 31) connected to a sensory nerve-fibre (B), in communication with a nervous centre or cell (C), which receives and transfers a nerve impulse through a nerve-fibrilla (D), to a nervous centre or cell (E), this forwarding an impulse through a motor nerve-fibre (L) connected with an internal or a peripheral organ (M). Nervous centres are generally connected with each other by nerve-fibrillæ; many of the latter together constituting the medullary or white nervous substance of the brain and spinal cord.

The co-ordinate movements and reflex acts may be effected in three different ways: 1. From internal organs through the sympathetic and cerebro-spinal system to voluntary muscles and sensitive surfaces. 2. From sensitive surfaces through the cerebro-spinal and sympathetic system to involuntary muscle-fibres and secreting organs. 3. From one internal organ to another, through the sympathetic nervous system.

It is owing to reflex action that the normal activities of the various organs, and their functions are maintained; as, for instance, the contact of food with the mucous membrane inciting the muscular coat of the intestines to peristaltic action; the gall-bladder discharging its contents into the duodenum during the process of digestion; the circulation of the blood becoming accelerated during digestion in order to facilitate glandular secretion. It is also due to reflex action that all kinds of disturbances of the physiological equilibrium are produced—such as convulsions in children from the irritation of undigested food; diminished or unnaturally increased appetite from any diseased part of the system; mental depression interfering with the digestive process or with secretion; terror, causing the pupil of the eye to dilate, or the action of the heart to become increased or diminished; in early pregnancy, the change in the condition of the uterine mucous membrane causing nausea and vomiting; fright, interfering with the menstrual function, or producing premature delivery or other disagreeable occurrences; local inflammation, arousing the general circulation, since increased heat augments reflex excitability.

Sympathy of one organ with another is the very foundation of the maintenance of life; for, if the action of one organ is increased and other organs act normally, the affected organ would soon become exhausted; but if the action of one is increased and others diminished through vital equilibrium, the system suffers less. So, on the other hand, the continued violent action of one part may affect others in such a degree as to cause cessation of all. These sympathetic functions are governed by reflex actions, and are further illustrated by the sympathy of the eye with the stomach, as in amaurosis, strabismus, and even hemiplegia from indigestion; in diarrhœa arising from coldness or wetness of the integument; in the sympathy between the stomach and the liver; of the intestines with the stomach and the brain; of the skin with the parts below; of the uterus or kidneys with the stomach; and, finally, in the circulation of the entire system, aroused by an inflammation or disturbance of one part.

Certain reflex actions may be controlled, or even prevented, by the will. External reflexes are illustrated by the contraction of the abdominal muscles in response to stimulation of the skin over the mammæ, sometimes called *abdominal reflex;* the *tickling abdominal reflex,* or contraction of abdominal muscles by stroking the skin over the side of the abdomen; the *plantar reflex,* by tickling the sole of the foot, producing movements of the toes, or of the toes and foot; the *pupil reflex,* being its dilation (the iris) on stimulation of the skin of the neck, or its contraction on exposure of the eye to light.

CHAPTER XV.

NERVE-FIBRES AND THEIR TERMINALS.

Nerve-fibres are cylindrical filaments, which generally run parallel in groups or tracts (A, B, Fig. 32). They are composed of an external sheath (1), axis-cylinder (3), and a medullary layer between these two (4). The sheath is a colorless, elastic, tubular membrane, which protects the internal parts of the fibre. The medullary layer is termed the *white substance of Schwann*. It is a transparent, oily substance, giving the fibre a white, shiny appearance. Physiologically, it confines the transmission of nerve-fibre currents (impulses) to the axis-cylinder, a pale, homogeneous granular, cylindrical cord in the center of the nerve-fibre. It differs from the medullary layer, which is nearly fluid in consistency, in being more solid and elastic, extending the entire length of the fibre, and of which it is the most essential element, since the currents pass through it.

Nerve-fibres vary in diameter from about 2 to 16 micromillimetres; the smallest being in the brain and spinal cord, the largest in the nerves of the periphery. In some parts of the nervous system they also differ in construction, some being *medullated* and others *non-medullated*, though not differing in function. The latter only have no medullary layer, are not so white therefore on account of its absence, and are found near the cells of gray substance of nervous centres and in nerve terminals. Again, a fibre may have a medullary layer through a portion of its course, but none at its termination, as in the cerebro-spinal nerves; or it may originate without this layer and after a short distance become medullated.

When nerve-fibres issue from bony cavities, they usually coalesce into bundles, which are held together by a thin connective tissue coat termed **Neurilemma** (A, or B, Fig. 32), derived from the dura mater and periosteum. Such a bundle of fibres is called a **Nerve** (A or B), and generally contains both sensory and motor fibres. These fibres do not branch or inosculate with each other in the nerve, though a fibre may separate from one nerve (2) and join another. Each fibre is continuous, and independent of every other,

Fig. 32.

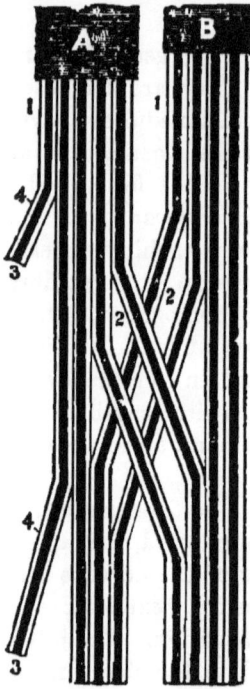

Diagram of two Nerves
and Their Fibres.

A, Nerve, the neurilem-
ma of which incloses five
nerve-fibres. B, Nerve,
the neurilemma of which
incloses four nerve-fibres.
1, Sheath of a nerve-
fibre. 2, Two nerve-fibres
separating from one nerve
and joining the other. 3,
Axis-cylinder in the cen-
tre of the nerve-fibre. 4,
Medullary layer, or white
substance of Schwann, be-
tween the external sheath
and axis-cylinder of the
nerve-fibre.

from its origin to its final termination; *i. e.*,
the sensory to the nervous centre-cells, the
motor to nervous plexuses of motor and
secreting organs. The neurilemma of the
nerve is endowed with sensitiveness by virtue of the cells of the neurilemma; it is,
therefore, excitable in any part of the
course of a nerve.

The white, marrow-like substance of the
brain and spinal cord is almost entirely
composed of nerve-fibrillæ, which in these
positions do not run in separate covered
bundles as nerves, as the cerebro-spinal
fibres do on the outside of the skull and
spine. Near the peripheral termination
the fibres change both in structure and
arrangement, since they divide and subdivide (Fig. 33), yet each branch of the subdivision retains all the original functions
of a nerve-fibre, and form many *plexuses*,
the fibrillæ of which supply the anatomical elements of the tissues. In the tissues,
especially in the skin, there are two of such
plexuses, a deep and a superficial one, the
latter being more closely arranged, and
most intimately associated with anatomical elements. At the place of subdivision
into a peripheral plexus, the nerve-fibres
are constricted (2, Fig. 33); each branch
that follows the constriction is nearly
equal in diameter to the fibre before subdivision. At the periphery the sensory
fibres terminate either in tactile corpuscles
(Fig. 34), Pacinian bodies (Fig. 35), or in
terminal bulbs (Fig. 36); the motor, in the
terminal plates (Fig. 37); the first three
being cell-like bodies. The various forms
of the terminal cells of nerve-fibres have
their respective special and separate secreting and other functions, in the various
organs and tissues of the body.

The **Tactile Corpuscle** (Fig. 34) is of an oval shape and situated in the papillæ of the skin, being most numerous near the tips of the fingers and in the soles of the feet. They are organs of the special sense of touch (*organa tactus*), most highly developed in those

Fig. 33.

Fig. 34.

Diagram of Nerve-fibre
Branching.

1, A nerve with three nerve-fibres.
2, The constriction of the nerve-
fibres at their point of branching.

Diagram of a Tactile
Corpuscle.

1, Nerve-fibre sheath. 2,
Axis-cylinder. 3, Termination of the axis-cylinder.

parts more numerously supplied with tactiles. The corpuscle consists of a central, transparent, gelatinous mass, surrounded by an envelope of connective tissue. It has many transverse elongated cells with nuclei, and is supplied by a nerve-fibre, which, on entering the corpuscle, loses its sheath (1). The sheath becomes amalgamated with the outer wall of the corpuscle. The medullary layer disappears also, so that the axis-cylinder (2) alone enters and terminates inside the corpuscle. The high degree of sensibility of a tactile corpuscle depends on its internal cells and its connection with the nerve-fibre. There are differently qualified degrees of the sense of touch. Thus, the modification of the sense of touch, by which certain qualities, such as softness, roughness, etc., of bodies are felt on very delicate contact, must not be confounded with ordinary sensitiveness to pressure and temperature, nor with pain, which pertains as such directly to the "Common Sense." It is said "as such" because pain has reference, too, to the sense of touch. Again, certain narcotic poisons will render a person insensible to pressure, pain, or temperature, without, however, affecting his

10

appreciation of the quality, for instance, of roughness in a body. In *analgesia* he becomes insensible to pain, but remains sensible to other impressions pertaining to the sense of touch. (*Vide* Senses.)

The **Pacinian Body** is oval-shaped (Fig. 35) and consists of connective tissue laminæ (2), inclosing a transparent, colorless, semifluid substance in its central portion. It is found on the peripheral terminations of some of the cerebral, spinal, and sympathetic nerves, in the subcutaneous connective tissue of the skin some distance beneath the derma, but not in it; also, either between or immediately beneath the epithelial cells of the mucous and serous membranes; in the conjunctiva and lips, and sensitive organs of the body. A single branch of a nerve-fibre (5) enters the Pacinian body. At the point of entrance into the body, the external sheath (4) is continuous with the external connective tissue lamina (1). The medullary layer of the fibre disappears, and only the axis-cylinder (3) enters the central portion, and continues to its extremity. The Pacinian bodies are also organs of touch (*organa tactus*), and are believed to be the organs on which sensation depends, such as pressure, pain, etc., which are common to all organs. Undoubtedly, the Pacinian bodies differ in size, structure, and function, more or less, in each organ. A good illustration is afforded in the peculiarity of the sense of touch after the *Taliacotian* operation (*Rhinoplasty*), where a portion of the skin of the forehead is partly turned downward (not cutting all the nerve-fibres) to make a new nose. When the nose itches the person scratches the forehead as the seat of sensation. This also evidences that external sensation lies in the organ, and not in the brain. (*Vide* chapter on The Senses.)

Fig. 35.

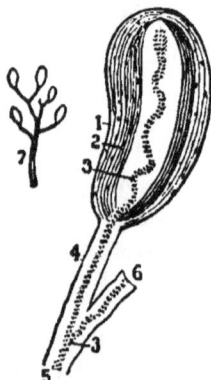

Diagram of a Pacinian Body.

1, Connective tissue layer. 2, Concentric layers of connective tissue. 3, Axis-cylinder. 4, Nerve-fibre sheath. 5, Nerve-fibre. 6, Branch of nerve-fibre. 7, Five Pacinian bodies and branches of a nerve-fibre.

The sense of touch is the most important and the most necessary for the vital economy, as without touch no external impression can be produced on any organ. The touch of the ray of light on the eye stimulates sight; the touch of the vibrating air (sound) stimulates hearing; the touch of substances to the tongue, taste; floating particles of matter in the air, smell; substances touching the œsoph agus, stomach, intestines, stimulate the secretions and peristaltic action; the touch of carbonic acid elements in the blood stimulates the motion of the heart, lungs, kidneys, liver, and skin to continual activity. When the sense of touch in the general economy is severely impaired, the person is dying, or, when entirely lost, dead.

The **Terminal Bulb** (Fig. 36) is either a round or slightly elongated, closed body of connective tissue, containing a fine granular substance, and transverse cells with nuclei. It is found within the papillæ of the mucous membrane of the tongue. A nerve-fibre (2) enters the papilla similarly to that of the Pacinian body, the sheath of the fibre continuing with the outer connective tissue layer (1). The sense of taste depends on the terminal bulbs, which differ from the tactile corpuscles in being smaller in size and of simpler structure. (*Vide* Glosso-pharyngeal Nerve.)

Fig. 36.

Fig. 37.

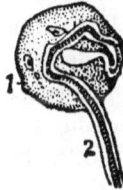

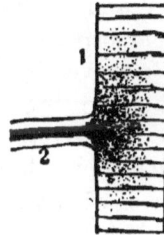

Diagram of a Terminal Bulb.   Diagram of a Terminal Plate.

Fig. 36.   1, Connective tissue layer. 2, Nerve-fibre.
Fig. 37.   1, Continuation of the nerve-fibre sheath with the sarcolemma. 2, Nerve-fibre, the axis-cylinder of which spreads out and forms the terminal plate in the muscular fibre.

The peripheral cells of nerve-fibre terminals and the cells of the nervous centres are two very important factors in nervous operations, while the nerve-fibres are simply conductors of the special nervous impulses between the peripheral and centre cells, the result being determined by the position and character of the terminal cell-organs. The external receiving organs are of various kinds, so as to be capable of appreciating the different kinds of stimulation presented, illustrated by those in the eye (cones) receiving the stimuli of light; in the ear (organ of Corti), sound; on the tongue (the terminal bulbs), taste; in the nose (olfactory membrane), smell; skin, finger ends in particular (tactile corpuscles), special touch; and all organs (Pacinian bodies), general touch.

The **Terminal Plate** (Fig. 37) is found in the muscular fibre. A nerve-fibre (2) penetrates the muscular fibre (1) at a right angle.

Its sheath becomes blended with the sarcolemma, the medullary layer disappearing, while the axis-cylinder enters the muscular fibre and expands into a *terminal plate*. (*Vide* chapter on Muscular Tissue.)

Nerve-fibres are the means of communication, connecting the nervous centres with the organs of sensation and of motion, and are endowed with *irritability*, by which impulses are transmitted throughout their length. The physiological attractive power of a stimulus for an impulse produced at the extremity of the nerve-fibre is similar to a magnetic bar, having attractive power only at its extremity.

The physiological *irritability of a nerve* is that property of becoming excited by the proper stimulus. Other tissues and organs possess a similar property if stimuli adapted to their physiological characters be applied. A gland, when stimulated, produces a secreted fluid; a muscle contracts, producing a movement of the parts to which it is attached. An impulse through a nerve-fibre brings about atomic motion (quivering), and, if sufficiently strong or prolonged, causes a change of the composition of molecules, and exhaustion in the axis-cylinder. This is proven by stimuli frequently applied in the same place, the irritabilty being soon weakened or destroyed. Nervous as well as muscular irritability becomes exhausted by continuous stimulation (excitation), but if at rest for a time their irritability will be partially or wholly restored, and if stimulus be applied may again become exhausted. The irritability and transmission of impulses of both sensory and motor nerve-fibres can be diminished or totally suspended by cold, compression, contusion, or laceration, if the injury is sufficient. After the nervous irritability is restored, the power of sensation and motion is re-established. The sensory (afferent) fibers transmit impulses inward to the nervous centres; the motor (efferent) fibres transmit impulses from the nervous centres to peripheral organs. As a rule, both the afferent and efferent nerve-fibres are associated in the same nerve and carry impulses in opposite directions without interfering with each other in their functions. No anatomical difference appears to exist between these fibres. In some nerves, however, the sensory and motor fibres are in separate bundles (nerves), like those of the facial nerve of the fifth cranial, being composed of sensory fibres which are distributed to the integument of the face and the mucous membrane of the mouth; while the branch of the same nerve distributed to the muscles of mastication consists of motor fibres. The fibres of a nerve separate near their terminal distribution.

The irritability of muscular fibres and that of motor nerve-fibres is distinct; which is proven by the action of woorara, which destroys the irritability of the nerve-fibres without affecting the muscular fibres. This accounts for two kinds of paralysis: 1. A nervous paralysis affecting the nerve-fibres only, the muscle being still contractable by direct stimulus. 2. A muscular paralysis, in which its fibres are directly affected, not the nerve.

The action by which sensory and motor nerve-currents are forwarded is very rapid; yet there is a short interregnum, and consequently it is not instantaneous. In sensation there are three successive stages—the reception of an impression by the sensitive nerve-terminal, the transmission of the impulse through the nerve-fibre, and the vital representation in the organ. (*Vide* chapter on Senses.)

The speed of transmission in motor fibres is about from 30 to 32 metres (about 104 feet) per second; in sensory, from 50 to 60 metres (about 196 feet). The transmission of impulses within the spinal cord differs very much—tactile impressions, 44 metres (about 144 feet); painful, 15 metres (about 49⅛ feet); and motor, 12 metres (about 39½ feet) per second.

It is an important fact that, should a nerve be divided in any part of its course, communication at once ceases, sensation can no longer be produced by impressions from without, and no voluntary contraction is incited in the muscles. A division of a nerve-fibre, or pressure sufficient to separate its axis-cylinder, though its sheath may yet remain intact, destroys the transmission of impulses. The division of a nerve-fibre, as well as its function, may be restored, if both ends at the point of separation remain in apposition; and even if a portion of a fibre is destroyed a reproduction of this part may take place. If a large nerve is divided, each one of its fibres again unites with its fellow, and not with others, so that the function of the entire nerve may be restored. The restoration of the normal function of a large nerve requires from three months to about two years, according to the age and condition of the individual. A division of small nerve branches produces a local anæsthesia in the immediate neighborhood, which continues for weeks or months after the healing of the traumatism.

Electrically, different actions can be produced by *direct* and *inverse* galvanic currents. If they traverse the nerve in the natural direction of its fibres, from origin towards termination, it is called *direct;* contrary, *inverse* current. When the nerve is fresh and irritable, or when the galvanic current is of sufficient intensity, a muscular contraction takes place at both the commencement and termination of

the current applied, whether direct or inverse. When exhausted
by the direct current, the nerve still responds to the inverse; or,
after exhaustion by the inverse, it still responds to the direct.
Nothing is so exciting to a nerve as the passage of direct and inverse
currents alternately in rapid succession. For such electric stimulus,
we have the faradic apparatus, in which momentary currents of
induced electricity are made to traverse the circuit in alternate
directions. (*Vide* chapter on Nervous Centres and Reflex.) .

CHAPTER XVI.

THE SPINAL COLUMN.

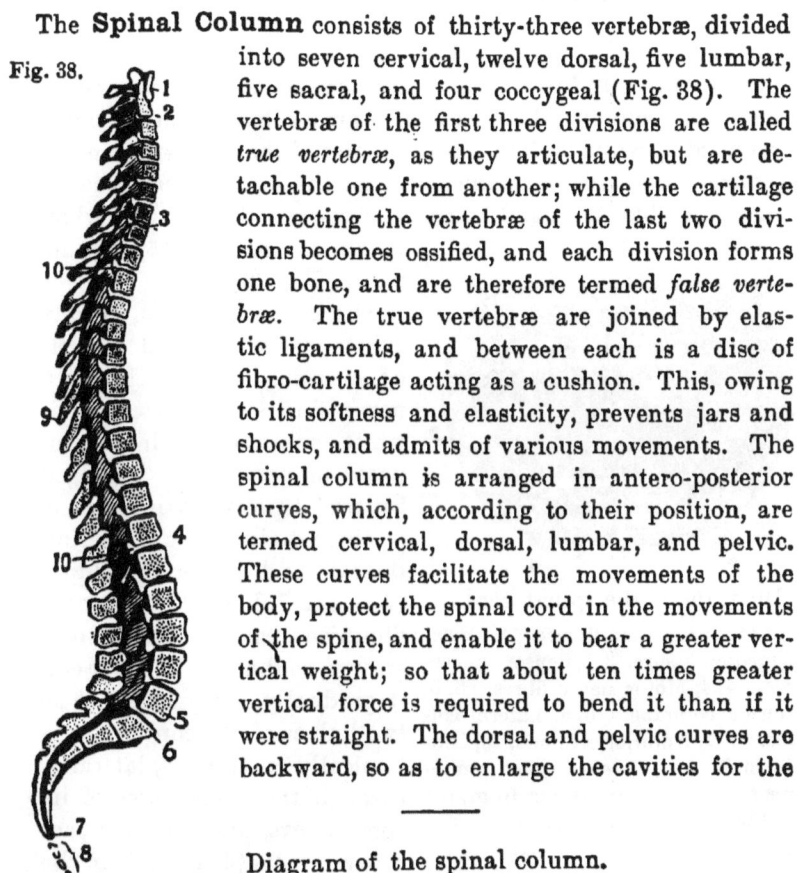

Fig. 38.

The **Spinal Column** consists of thirty-three vertebræ, divided into seven cervical, twelve dorsal, five lumbar, five sacral, and four coccygeal (Fig. 38). The vertebræ of the first three divisions are called *true vertebræ*, as they articulate, but are detachable one from another; while the cartilage connecting the vertebræ of the last two divisions becomes ossified, and each division forms one bone, and are therefore termed *false vertebræ*. The true vertebræ are joined by elastic ligaments, and between each is a disc of fibro-cartilage acting as a cushion. This, owing to its softness and elasticity, prevents jars and shocks, and admits of various movements. The spinal column is arranged in antero-posterior curves, which, according to their position, are termed cervical, dorsal, lumbar, and pelvic. These curves facilitate the movements of the body, protect the spinal cord in the movements of the spine, and enable it to bear a greater vertical weight; so that about ten times greater vertical force is required to bend it than if it were straight. The dorsal and pelvic curves are backward, so as to enlarge the cavities for the

Diagram of the spinal column.

1, First cervical vertebra, or atlas. 2, Second cervical, or axis. 3, Seventh, or last cervical vertebra. 2 to 3, Cervical curve forward. 3 to 4, twelve dorsal vertebræ (curve backward). 4 to 5, Five lumbar vertebræ (curve forward). 6 to 7, Five sacral vertebræ, (the pelvic curve backward). 8, Four coccygeal vertebræ. 9, Spinous processes projecting from the vertebræ. 10, 10, Foramina through which the spinal nerve-fibres issue, formed into nerves. 2 to 6, Roundish-square projections anteriorly, termed BODIES of the vertebræ.

thoracic and pelvic organs. The cervical and lumbar curves and the coccyx have a position forward, in order to support the parts anterior and above them.

THE SPINAL CORD.

The **Spinal Cord** is a long, cylindrical, nervous mass inclosed in the spinal column. It consists of a gray substance in its center, surrounded by white matter. In the center of the gray substance is the *central canal,* which commences near the *calamus scriptorius* in the medulla oblongata, and extends downward through the whole length of the cord (6, Fig. 26; and 8, Fig. 39).

Fig. 39.

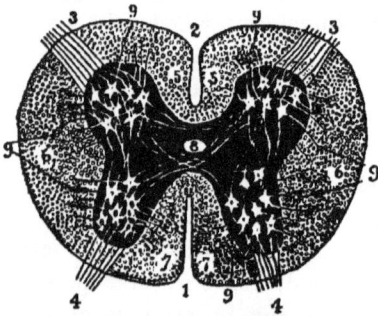

Diagram of the spinal cord.

1, Anterior median fissure. 2, Posterior median fissure. 3, 3, Sensory nerve-fibres. 4, 4, Motor nerve-fibres. 5, 5, Posterior columns. 6, 6, Lateral columns. 7, 7, Anterior columns. 8, Central canal. 9, 9, Cerebro-spinal nerve-fibres entering to, or issuing from, the spinal nervous-centre-cells.

The internal portion, or gray substance, of the cord consists of nervous centre-cells and nerve-fibrillæ; the external, or white substance, being composed of nerve-fibres, arranged by two depressions into tracts of three divisions—an anterior, lateral, and posterior column (5, 6, and 7, Fig. 39). The cord diminishes as the nerve-fibres emerge, so that at the lower end it gradually disappears. The two halves of the cord are anatomically continuous, laterally, through the nerve-fibres of its commissure, and, therefore, act only as a single organ. The anterior nerve-roots (4) issue from the cells of the gray substance of the anterior horns, and convey impulses outwardly for the motion of muscles, for secreting and excreting organs, likewise for vaso-motor, vaso-dilator, and trophic functions; while the posterior nerve-roots (3) are connected with the posterior horns of gray matter, and convey sensory impulses inward to the cord and brain. The cells of the anterior horns (motor) are large with multipolar processes; while the cells in the posterior horns (sensory) are smaller and spindle-shaped. The nerve-fibres of the spinal nerves and the fibres of the sympathetic system freely communicate with each other in the neck, thorax, and abdomen.

The spinal nerve-fibres pass from the cord through the intervertebral foramina (10, 10, Fig. 38), formed into thirty-one pairs of nerves, which are finally distributed to the various parts of the body. Eight pairs are cervical; twelve are dorsal; five, lumbar; five, sacral; and one coccygeal. Each nerve has two roots—an anterior and a posterior root. The anterior root-fibres are motor; the posterior, sensory. The latter, before entering the cord, pass through a ganglion (10, 10, Fig. 40) consisting of gray matter with bipolar cells.

Fig. 40.

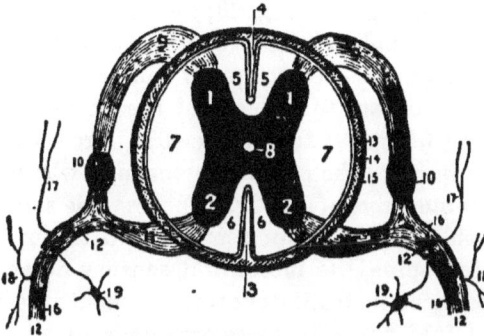

Diagram of spinal nerves.

1, 1, Posterior horns of the gray matter. 2, 2, Anterior horns of the gray matter. 3, Anterior median fissure. 4, Posterior median fissure. 5, 5, Posterior columns. 6, 6, Anterior columns. 7, 7, Lateral columns. 8, Central canal. 9, 9, Sensory nerve-fibres. 10, 10, Ganglia on the sensory nerve-fibres. 11, 11, Motor nerve-fibres. 12, 12, Spinal nerves. 13, Pia mater membrane. 14, Arachnoid membrane. 15, Dura mater membrane. 16, 16, Neurilemma. 17, 17, Nerve-fibres supplying the muscles and skin of back. 18, 18, Nerve-fibres supplying the muscles, skin, and sides of the body. 19, 19, Sympathetic ganglia .

The *Law of Waller* shows that the nutritional activity of a nerve-fibre is in the direction of its physiological activity, and that its nutrition is influenced by the nerve-cell with which the fibre is connected. If, therefore, the posterior root (9) be divided between the cord and the ganglion (10), the fibres in connection with the ganglion remain unaffected, while the interior ends degenerate. If the anterior root (11) of a spinal nerve be divided, in a few days the ends of fibres, cut off from the cord, degenerate, while the ends attached to the cord are still nourished. The effect of this degeneration on the motor fibres extends through the nerve to its final terminations. In degeneration of the fibres the medullary substance of *Schwann* (4, Fig. 32) disappears, and the axis-cylinder disintegrates, forming drops of fatty matter.

The brain motor fibres that supply the body and limbs pass from one hemisphere to the other at the anterior pyramids of the medulla oblongata, where most of them cross the median line (M, Fig. 28), to the lateral columns of the spinal cord. Here they enter the multipolar cells in the anterior gray horns, which then give off the anterior spinal nerve-fibres (motor) (4, Fig. 39; and 11, Fig. 40). Some of the brain motor fibres do not cross at the medulla, but pass to the anterior pyramidal tract of the medulla, and continue to the anterior column of the cord (7, Fig. 39; and 6, Fig. 40), then passing to the sympathetic ganglionic cells of trophic (nutritive) centres. The impulses through the cerebro-spinal sensory nerve-fibres pass from the peripheral organs to the cells of the posterior gray horns of the spinal cord (3, Fig. 39; and 9, Fig. 40). From these cells the fibres proceed upward to the medulla oblongata, where they cross, similarly to the motor fibres (M, Fig. 28), from one side of the cord to the opposite side, to the brain centres. On the outside of the bony cavity of the spine, the motor and sensory nerve-fibres proceed together, coalesced in trunks (nerves), without interfering with each other. A spinal nerve is, therefore, made up of sensory and motor nerve-fibres, which transmit the impulses necessary for sensation and motion.

By this it is proven that the spinal cord is the exclusive organ of communication between the brain, the body and limbs. It also sends out nerve-fibres from centres of its own, to supply the integument and muscles of the neck, trunk, and extremities, and contains reflex centres for the sphincter, or constrictor muscles of the organs of evacuation. The small intestines, cæcum, and colon are supplied exclusively with nerves from the abdominal plexus of the sympathetic system; while the lower portion of the rectum, mucous membrane, and sphincter muscles receive branches from the sacral plexus of spinal nerves. The retention and discharge of feces are effected by two sets of muscular fibres, which are regulated by reflex influence through the spinal cord. One of these muscles is the *sphincter ani*, which keeps the orifice of the anus closed, while the *levator ani* and circular muscular fibres of the rectum, open it. Any external irritation to the sphincter ani increases the contraction of its fibres, closing the anus more completely. Its habitual closure is entirely involuntary during sleep and in waking hours, depending upon the reflex action of the spinal cord; but as soon as the rectum is distended to a certain point by feces, the nervous action changes. The irritation then produced upon the mucous membrane of the rectum sends forth an impulse through its sensitive nerve-fibres to the spinal cord, which reflexes an impulse of relaxation to the sphincter; while, simultaneously, the levator ani draws the borders of the relaxed orifice upward and outward, and the feces are expelled by the contraction of the muscular fibres of the rectum itself. Resistance to evacuation by the sphincter may be voluntarily prolonged; but in time the reflex impulse to evacuate becomes so urgent and irresistible that it cannot be controlled by the will, and the discharge occurs involuntarily, simply by reflex action of the spinal cord. If the lower portion of the cord be disorganized by lesion or paralysis, it then loses its power to reflex, and the sphincter a n i becomes permanently relaxed, and the feces are evacuated as fast as they descend into the rectum, without the knowledge or control of the patient.

The supply of blood to the spinal cord is derived from the vertebral arteries, and from branches of the intercostal, lumbar, and sacral arteries. At the parts where the greatest number of nerve-fibres emanate, the spine is enlarged. Three enlargements especially deserve notice—the cervical, dorsal, and lumbar. The cervical is the principal source of nerves for thoracic organs and respiratory muscles; the dorsal is the source of nerves for the upper extremities, and partly for the thoracic and abdominal organs and muscles; while the lumbar enlargement is the principal source of nerves for the pelvic organs and the lower extremities. The spinal gray matter contains also centres independent of the brain in their reflex functions, viz.: 1. A *cilio-spinal* reflex centre, situated between the sixth cervical and third dorsal vertebræ; being the normal dilating centre of the iris—*i. e.*, for dilatation of the pupil. 2. A *genito-spinal* reflex centre, situated about the middle lumbar region of the cord, reflexing impulses to the generative organs; injury to this centre is followed by loss of virile power. 3. In the same region is the reflex centre for parturition pains in the female. 4. In the upper lumbar region are the *ano-spinal* and *vesico-spinal* centres, reflexing motor impulses to the sphincter a n i and the sphinter v e s i c æ. Loss of reflex at these two centres causes involuntary evacuation of the rectum and complete paralysis of the bladder, the urine constantly escaping in d r i b b l e t s, though the bladder may remain partially filled.

Section of the cord between the third and fourth cervical vertebræ paralyzes the phrenic nerve and arrests the action of the diaphragm; section between the fifth and seventh p a r a l y z e s the posterior thoracic nerve and thoracic muscles; section from the first to the ninth dorsal paralyzes the intercostal muscles; section from the seventh to the twelfth dorsal, the abdominal muscles.

The spinal cord has principally a *three-fold function:* 1. It serves as a medium for conducting impulses through nerve-fibres from the brain to the peripheral organs, and *vice versa.* 2. In the normal condition many spinal centres are subservient to centres in the brain, so that the cord reflexes may be inhibited or restrained, more or less, especially over movements by spinal-reflex on parts that can be moved by the will. 3. In case the reflex action of the cord be on a portion of the body which is invariably involuntary, then such action cannot be inhibited by the will. This is illustrated by the cilio-spinal centre, acting on the iris, dilating the pupil; by the parturition centre, acting on the uterus during labor pains. If the spine be separated from the brain, some centres in the cord are still capable of action, such as those which by reflex effect the closure and dilatation of the sphincter muscles, and muscular contraction, showing a power of reflex action by the cord independent of the brain.

The **Urinary Bladder** is both a reservoir and an organ of evacuation. It is protected at the commencement of the urethra by a circular bundle of muscular fibres, known as the *sphincter vesicæ,* and is supplied by nerve-fibres from the hypogastric and sacral plexuses. Those from the hypogastric are distributed to the upper part of the neck of the bladder; those from the sacral, to the lower part. The tonic contraction of the vesical sphincter during health is continuous and involuntary, like that of the sphincter ani. When micturition occurs the sphincter vesicæ relaxes, while the muscular coat of the bladder contracts to expel its contents. A relaxation of this sphincter may also take place by voluntary impulse, acting on the spinal reflex centre of micturition; but the subsequent contraction of the bladder continues without voluntary effort. Lesion of the hypogastric plexus may paralyze the upper part of the neck and bladder; lesion of the sacral plexus may paralyze the lower part. The principal reflex impulse for contraction of the bladder emanates from the spinal cord through the sacral nerves. In inflammation of the bladder, or neighboring parts, its sensitiveness is increased, and the reflex impulse to urinate cannot be resisted. Increased acidity of the urine produces a similar result. Paralysis of the bladder causes its permanent distension, followed by a continuous, passive, but incomplete, discharge.

Normally, the reflex actions of the spinal cord are chiefly of an expulsive kind — to force out the contents of the various cavities, such as defecation and urination, ejaculatio seminis, and parturition, over some of which the will has very little or no control.

The decussation, or crossing, of both motor and sensory nerve-fibre tracts is at the medulla oblongata. A lesion in a certain portion in a lateral half of the brain above the medulla oblongata causes paralysis on the opposite side of the body; but if the lesion is below the medulla oblongata on the lateral half of the spinal cord, then the paralysis is on the same side of the body. Paralysis of one-half of the body is termed *hemiplegia*. A lesion involving the middle or upper entire portion of the cord transversely, induces a paralysis called *paraplegia*, or palsy transversely; *i. e.*, paralyzing legs and thighs and lower part of body, including the bladder and rectum. The communication between them and the brain being arrested, sensation is no longer experienced in said parts, nor can the will produce contraction of muscles below the lesion. The seat of lesion in any part of the spinal cord is determined by the line at which paralysis of sensation, motion, or both, manifest themselves in the external parts—if in the lumbar portion, then the legs and pelvic region are paralyzed; if in the dorsal, all below the thorax and abdomen; and if in the middle cervical region, then both the upper and lower extremities, chest, and intercostal muscles. A lesion, injury, or disease, though very slight, if just above the middle cervical part of the cord, is a dangerous affair, because it cuts off the reflex motor impulses to the respiratory muscles, producing death by stopping respiration. The phrenic nerve arises at the third, fourth, and fifth cervical, and supplies the diaphragm.

The physiological effect of strychnine is directly upon the cord, producing rapid irritation, and augmenting its excitability for reflex action, so that the slightest external stimulus causes excessive muscular action and convulsive movements. A wound affecting peripheral nerves may produce tetanic spasms or convulsive actions by reflex of the cord, and when the increased irritability of the cord is once established, it is easily excited by an external cause, which may be so slight that a healthy person would not perceive it. During such an affection, startling movements or sounds of any kind, or even a current of air upon the skin may, by reflex of the cord, throw the muscular system into severe spasmodic action. The spinal cord as a nervous centre is apparent in such convulsive diseases as hydrophobia and tetanus, over which the will has little influence.

In **Hydrophobia,** not only the spinal cord, but the medulla oblongata and some of the cerebral ganglia, are involved. The effect is brought about by blood-poisoning; but, strangely enough, the rest of the brain completely escapes its influences. The stimulus most effective in hydrophobic convulsions is through either one or both of two senses—the *sight* or *sound* of water causing a paroxysm, and any attempt to taste it increasing the severity.

In **Tetanus** there is a similar excitable state of the spinal cord and medulla oblongata, but it does not involve the ganglia of any of the five external senses, except that of touch. The cause of tetanus may be altogether internal, as in the idiopathic form—a condition resembling that which may be artificially induced by strychnine. It may also arise from traumatism, as a lacerated wound, the irritation of the peripheral injured nerve terminal being propagated to nervous centres, establishing the excitable condition; and, once established, the removal of the original cause by amputating the injured limb seldom does any more good, because the slightest impression upon almost any part of the body is sufficient to excite the tetanic spasm.

Epilepsy, or convulsive movements with temporary suspension of the functions of the cerebrum, generally results from internal organic functional derangement, which, when established, causes also an excitable condition and disorder of the nervous centres. It may result from irritation proceeding from local causes, like the convolsions of teething, and cease when the irritation is removed.

Convulsions which end fatally usually act by suspending the respiratory movements, the muscles which effect respiration becoming fixed by the spasms, so that suffocation occurs as completely as if the air passages were ligated.

Hysteria is a peculiarly excitable state, in which every kind of convulsion may be simulated, although there is seldom present a sufficient indication of any positive disease. One form of convulsion after another often occurs at short intervals, and with wonderful variety. Hysteria may be likened to convulsions from local irritation, such as hysteric spasms at the catamenial period, especially if menstrual flux be deficient. A diagnostic feature distinguishing hysteric from epileptic convulsions is that in the former the patient does not bite the tongue, the centre at which the hypoglossal nerve-fibres originate seemingly remaining unaffected.

The membranes covering the brain and spinal cord are termed *meninges,* and an inflammation thereof *meningitis,* which, if on the posterior portion of the cord, causes great pain; if on the anterior portion, however, convulsive movements without pain ensue. Usually the morbid action extends to the anterior and posterior membranes and columns, thereby causing severe pain and disturbing motion.

CHAPTER XVII.

THE SYMPATHETIC SYSTEM.

The **Sympathetic Nervous System** consists of ganglia of gray matter (forming nervous centres), and slender nerve-fibrillæ issuing therefrom. It is not an entirely independent system. Its ganglia are not only connected with each other, but also communicate with the cerebro-spinal system by nerve-fibrillæ. Along each side of the spinal column the ganglia lie in the form of a chain-like continuation, though in other parts of the body they are scattered irregularly. A **Ganglion** is a discrete encapsulated portion of gray matter in the course of, and connected with, nerve-fibres, as the *Gasserian* ganglion of the fifth nerve, or the *Otic (Arnold's)* ganglion below the foramen ovale. Each ganglion contains nervous centre-cells with multipolar processes (C, Fig. 30). The nerve-fibrillæ issuing therefrom freely intercommunicate with each other and with the chief ganglia, and are of two kinds — medullated and non-medullated. The medullated are derived from the cerebro-spinal system, and pass through the sympathetic ganglia, supplying the vascular and secreting organs. The non-medullated fibres are destitute of the white substance of Schwann. They consist of a gray, gelatinous substance, and originate in the sympathetic ganglia. Through these fibres, sensory or motor, the sympathetic impulses are transmitted, thereby establishing that sympathy, greater or less, existing between all organs of the body. Its nerve-fibrillæ are distributed to the walls of the blood and lymphatic vessels, interlacing them as ivy entwines the tree. They accompany these vessels throughout their ramifications and to their utmost terminations. They are also distributed to the organs of the head, neck, trunk, and extremities.

The more delicate nervous centres of the brain and spinal cord are wonderfully protected from over-irritation. This protection lies in the modifying power of the sympathetic ganglia placed on the tract of sensory nerve-fibres of each spinal nerve (1, 1, Fig. 41). The sensory fibres to the cerebral convolutions pass through the optic thalamus (D, Fig. 28).

The sympathetic nervous system is most largely distributed within the chest and abdomen, connecting the organs of nutrition, such as the secretory, digestory, absorbent, and the circulatory appara-

Fig. 41.

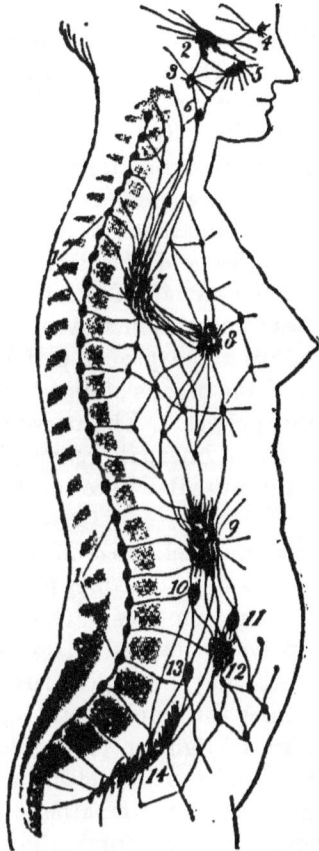

Diagram of the sympathetic system.

1, Ganglia on sensory nerve-roots of right side of spinal column. 2, Gasserian ganglion. 3, Otic ganglion. 4, Ophthalmic ganglion. 5, Sphenopalatine ganglion. 6, Submaxillary ganglion. 7, Deep cardiac plexus. 8, Superficial cardiac plexus. 9, Semilunar-ganglion plexus. 10, Renal plexus. 11, Aortic plexus. 12, Epigastric or Solar plexus. 13, Spermatic plexus. 14, Hypogastric plexus.

tuses. It also presides over the functions of the internal organs. Besides its many ganglia of varying sizes, it forms more than fifty **nervous plexuses,** such as the *aortic, cardiac, cervical, epigastric,*

11

hepatic, hypogastric, lumbar, pulmonary, renal, sacral, solar, spermatic, and many others.

The peristaltic action of the œsophagus, stomach, and intestines is by reflex, through the motor fibres, from the nervous centres in the medulla oblongata. The vaso-motor centres of the medulla reflex impulses for dilatation and contraction to the vascular system, in a similar manner. These nerve-fibres pass out through the spinal foramina with the spinal nerves (Fig. 41), but soon leave them to enter sympathetic ganglia (19, Fig. 40). This accounts for the harmonious co-ordination of functions of the medulla oblongata with those of the sympathetic system and of the organs of the body.

The peculiarity of the sympathetic nervous system is the small diameter of its nerve-fibres and the difference in function from that of the cerebro-spinal system. The latter system manifests its most striking properties in sensation and motion; while the former preponderates in regulating the proper performance of the functions of co-ordination of the organs of digestion, absorption, assimilation, circulation, and growth. The nerve-fibres of the sympathetic are also distributed to internal organs, glands, mucous membranes and their internal muscular fibres, such as are not controlled by the will.

Physiologically, the sympathetic nervous centres, or ganglia, are endowed with sensitiveness and the power of conducting, reflecting, augmenting, and inhibiting impulses. The ganglia of the sympathetic system are sensitive to impressions, and are capable of perceiving irritations, which may even give rise to acute pain. The reflex action, which causes the pupil of the eye to contract or expand owing to variation of light, is propagated through the ophthalmic ganglion; therefore, the iris acting somewhat sluggishly, indicates the intervention of the sympathetic system, which in all instances requires more continued stimulation and is less rapid in activity, but of longer duration, than those of the cerebro-spinal system.

CHAPTER XVIII.

THE SENSES.

A **Sense** is the capacity or power of the animal for a particular species of what are termed *sensations*. The special portions of the organism endowed with the simple power of reacting to appropriate stimuli so as to produce these particular groups of sensations, are called **Sense-organs**. Sense, therefore, is an organic faculty, or one that pertains not to the living vital principle (*soul**) alone, but to the composite, or that which is made up of soul and matter conjoined. As sensation is the proper operation of the animal inasmuch as it is animal, and as the animal is not soul alone, but a compound of body and soul, so the faculty of sense pertains to the composite, and is an organic faculty. Sensation is in itself simple, for there is no one who does not see that a sensation, considered as such, cannot be divided. We have said "as such"; for, although the animal under certain circumstances may have only a portion of the sensibility it once enjoyed, on account of the material organ, yet that which it now has considered as sensation cannot be divided. In the sensation itself, however, we can distinguish the formal element from the material one: viz., the vital representation and the nerve vibrations, which latter element is extended and divisible. Sensation, therefore, is in itself simple, unextended, and indivisible, but, considered in reference to the organ and nerve vibrations, extended and divisible.

Sight, Hearing, Smell, Taste, and **Touch** constitute the five *external senses*, by and through which the perfect animal becomes cognizant of external objects, of their physical appearance, qualities, quantity, etc., while at the same time it is endowed with four *internal* senses—the *common sense, imagination, estimative faculty* or instinct, and *sensitive memory*. As to the existence and reality of the five external senses, no one can reasonably call them in question; but as to the four internal, equally undeniable though they be, yet unaccustomed as we are to hear them even mentioned, or if we do, to hear them spoken of in reference to man alone, and then often either considered as powers belonging only to the mind,

---

* See Note, page 129.

or else some of them, such as the sensitive memory and the common sense, confounded with the inorganic faculties of intellectual memory the understanding, it will not be out of place here to lay down a few clear notions on this point. We will take each of the internal senses in order, give an idea of its nature, and the proof of its existence, together with necessary remarks:

I. The **Common Sense.**—Every one has experienced that he not only perceives external objects proper of each of the external senses, but also feels the very sensations themselves as they are perfected in the external organs, and between them to discriminate one from the other. In other words, he not only perceives, or "senses" exterior objects by the external senses, but also feels *the very acts* by which they are perceived or "sensed," and distinguishes between them. Now, this cannot be done by the external senses, as these perceive only those objects proper of them. There is need, then, of some faculty pertaining to and not above the order of sensibility, which perceives these external sensations and discriminates between them. Thus, for instance, we distinguish a white thing from a sweet thing, and the sensation of whiteness from the sensation of sweetness; but as the sense of sight and the sense of taste (which are the senses here concerned) know only their own proper objects, it follows that there must be some other sense which perceives and distinguishes between them. If it be said that the intellect does this, it is answered that the same thing happens in the brute animal, which has no intellect. Is it not according to experience that a dog or a horse, for instance, raises its ears to hear better, thus showing that it *feels* the very external sensation itself? Are not animals sometimes so affected by music as to give full evidence of it by cries and other signs? Again, who can deny that the brute animal *feels* that it sees, *feels* that it tastes, *feels* that it smells? Now, the eye cannot distinguish between color and taste, the tongue cannot distinguish between taste and smell. The same likewise distinguishes between the objects of the different senses. Thus, a dog seeing an object at a distance which appears to its sight as a piece of meat (a painting), on approaching and smelling the same, quickly rejects it. There must then be an *organic* faculty which receives, feels, and distinguishes all these impressions with the objects (stimuli) that excite them, and this faculty is called *the common sense,* or by some *the internal sense.* That the animal has not understanding, and therefore no intellect, is admitted by every one of real philosophic acumen. The *discrimination,* be it remarked, which is spoken of above, must not be confounded with judgment properly so called, for it is nothing more than a true though sensible *apprecia-tion* of external sensations by means of the common sense. This operation of the common sense, in reference to external sensations, is sometimes called *conscience*, the proper meaning of which we will speak of in its proper place.

Having shown that the brute animal as well as man possesses the internal or common sense, and supposing here that in the brute animal there is no faculty that goes beyond the purely organic, it clearly follows that this common sense must be a faculty operating through an organ, laying aside all controversy as to this organ's location, and contenting ourselves with having shown the necessity of its existence. In like manner it clearly follows that, if in the sequel it can be shown that the perfect animal possesses the other three internal faculties above mentioned, namely *the imagination, the estimative faculty*, and *sensitive memory*, we have at the same time shown the existence of three other *internal organs*.

Before coming to the *imagination*, which is the second internal faculty in the order of nature, we will add something on *external* and *internal sensation*, and on *conscience*, for the better understanding of what has been already said. We have spoken of *external* sensations, which implies that there are also *internal* sensations. If sensation is defined, as it truly is, *a vital organic representation*, it is very evident from the very definition of sensation that, since there is at least one internal sense, as we have just now shown, and that this sense requires an organ, there must be *internal* as well as *external* organs quite distinct among themselves, and a dependency of the former on the latter. This dependency and connection is proved both by experience and by the very object of the internal sense as stated above. There is a great mistake made by some in supposing that external sensation—for instance, that of sight—is produced in the brain. This error arises from the circumstance of finding in the case of the eye (and what is said of the eye is equally true of the other external senses) a sensory nervous centre in the brain, with which the optic organ is connected. Now, the case is far otherwise, for it is according to experience that we see with the eye, hear with the ear, etc., and not with the brain. Sensation, therefore, if experience is to stand, takes place in the organs themselves. A pain in the foot is really there, and not in the brain. Besides, if the sensation of hearing, for instance, is in the brain, why is the ear so wonderfully formed, since a simple system of nerves

would suffice for transmitting the impression made by the sounding body to the brain. The same can be said of the other sense-
organs. This opinion, holding sensation to be produced in the
brain, comes often from the false supposition that the soul (vital
principle) does not reside in the whole body but in the brain only.
How erroneous such an opinion is, will be rendered clear if we
keep in mind that the soul is the principle of life and must be present by its essence wherever there is life and vital power. Now,
there is certainly life and vital power, not only in the whole body,
but in each single part of it. Therefore, the soul must be present by
essence, not only to the whole body, but to each single part of
it. If it be objected that, having destroyed communication between the external organ and the brain, sensation is destroyed,
and consequently that sensation is produced in the brain, the
objection may be answered thus: 1. That, having destroyed the
organ, sensation also is destroyed, and therefore, sensation is
produced *in the organ*. 2. The necessity of nervous centres in
the brain is not denied, for it is according to the most common
physiological experience that, separated from these centres, the
organs are inadequate to the production of sensations. That
exterior sensations, however, are produced in the brain is denied.
As the common sense is a faculty which is concerned with what
is popularly termed *consciousness*, it is proper to state in this
place what is strictly meant by the consciousness of a sensation,
which even the brute animal possesses in common with man.
Consciousness, *psychologically* considered, is properly and strictly
defined as that power of the *rational soul* of returning *completely upon itself*, by which the intellect not only perceives the
acts of the other faculties but its own acts also, and recog-
nizes itself as the subject of them. Now, the reason why the
rational soul alone, which is of man, has such a capability,
is because of the inorganic nature of that act, which is in
itself independent of matter and performed without the in-
strumentality of any organ. Hence to a principle, which
although simple in itself is yet essentially dependent on matter and cannot act but conjointly with an extended organ as is
the brute soul, this complete return on itself cannot take place on
account of that matter to which it is united being extended and its
parts consequently existing one out of the other. This term *consciousness* is properly and strictly applied only to certain intellectual acts of the rational soul, and does not pertain to the order of
sensibility. As, however, there is no other English word so well

adapted to convey the meaning of that operation by which the common sense takes cognizance of, and appreciates external sensations, we can by giving notice of the fact so as to preclude all misapprehension, apply the term to this act also.

II. The **Imagination** is the faculty of representing corporeal things that are absent. This is to be understood in the sense, that the operation of this faculty does not represent its object inasmuch as it is really present, and not in the sense that it does not operate *except* in the absence of the object; for in every impression of a sensible exterior object on our senses, as, for instance, an object of sight, both the act of vision of that object, and the act of the imagination of the same object, are brought into play at one and the same time. To prove this fully would detain us much too long, and involve much metaphysical reasoning. Suffice it to say, that the acts of the intellect while the human soul is united with the body, are always accompanied by an act of the imagination—a thing which is confirmed by experience. The imagination is a faculty not only for representing objects of the sight that are absent, but also for representing objects of the other external senses also in their absence. This is true, likewise, in regard to the objects of the common sense which it also reproduces in their absence. Thus, after having seen a beautiful picture, we can in the quiet of our room conjure up its representation. Having listened to some delightful music, we can in like manner enjoy it again in its absence. Dreams also take their rise and are seated in this faculty of imagination, which, when we are asleep, brings before us a representation of things that may be long since past. Now, no one can deny this power to man; and yet the brute animal, too, has this same power, but not in so high or excellent a degree as man, on account of the union and connection of this faculty with man's intellect. To show that animals possess this faculty of imagination, does not the cat, for instance, on hearing the sound of mice go in search of them, their phantasm in him being awakened by the sound it now hears? Again, do not dogs, cows, and horses give evident signs that they sometimes dream?

This faculty, according to what we have laid down above, is *organic*; and consequently, its organ, since it is not found among the external, must be found among the internal organs.

III. The **Estimative Faculty** is a sense endowed with the power of estimating or valuing objects as good or hurtful for appetite, or of apprehending an object as fit or unfit for the needs of animal

life. This is the power often called instinct. Thus the chicken
flees from the hawk, and the sheep from the wolf; not because the
hawk or wolf has anything repulsive in outward appearance,
since they do not flee from much more repulsive-looking animals.
It cannot, moreover, be owing to anything the chicken or the sheep
has learned by experience, for even the young chicken and the
lamb, although they never saw either hawk or wolf, will flee from
them at first sight; the chicken from the hawk, and the lamb from
the wolf. Again, the dog goes in search of grass in time of sickness,
having the instinct of its fitness for nature at that particular time.
This power is organic from what has been said above, and its organ
is internal. The same power is found also in man, which, however,
on account of its higher grade in him is called *cogitative power*.

IV. The **Sensitive Memory** is a faculty of the perfect animal for retaining those things apprehended by the *sensitive faculty*.
Thus, a horse remembers the stable on account of the food which it
receives there, and the dog the stick with which it was struck, on
account of the pain. This goes to show that this memory is of
things apprehended as agreeable or unpleasant, useful or hurtful to
nature, and therefore of things as we have said, apprehended by the
estimative faculty. This faculty is found in man; but on account of
its conjunction with the intellective soul is called *reminiscence*. It
must be organic from what has been said, and its organ internal.

Having given, as we think, a sufficiently clear account of the internal senses as we proposed to do, we come now to an explanation
of the terms used in defining sensation, to the elements of sensation
and the different sciences under which each of these elements fall.

Sensation, as we defined it above, namely, in the proper philosophical sense, is *vital, organic representation*. It is *vital*, in contradistinction to representations merely physical, which, for instance,
happen in the impression of the image of an object on the retina of
the eye; for whether the animal be dead or alive, this image is
equally impressed. It is *organic*, in contradistinction to those representations produced without the intrumentality of organs, such
as the intellectual or mental ones. Sensation, therefore, comprehends
in its essence three distinct elements: the *object*, the *organ*, and the
vital principle. On the part of the object, an immediate action on
the faculty through the organ, by the physical outward world in
the case of *external*, and a mediate one in the case of *internal* sensations, is required; the investigation of the nature of which constitutes the object of *physics* and *metaphysics*. The investigation of

the material character of the process within the organism, which goes before, accompanies, and follows the vital representation, or as they now say of the function of the organ of sensation, makes up the subject-matter of *physiology*; while the investigation of the sensitive faculty and operation itself belongs to *psychology*. With the first of these we have nothing to do here, and therefore relegate it to its proper domain. The third element we leave to the psychologist, and reserve the second for ourselves, in so far as *external* sensation alone is concerned.

A nervous apparatus is simply a nerve-fibre which connects two cells. Consequently, for effecting the merely material action within the organism three structures are necessary: a *peripheral end organ*, a *sensory nerve-fibre*, and a *nervous centre-cell*. The peripheral end organ receives stimuli, which, acting on the sensitive faculty through its organ, connected by the sensory nerve-fibre with the center-cell in the cerebrum, produce the vital external representation (sensation). The different senses, as said before, require special nerve centres of sensory areas in the cerebrum though these are not the very organs of exterior sensations as we have before shown. Thus sight requires a centre in the occipital lobe; hearing, one in the temporo-sphenoidal gyrus; smell, one in the lower portion of the temporal lobe; taste, one near that of smell; touch, one in areas of the limbic lobe, *i. e.*, convolutions surrounding the corpus callosum such as the gyrus fornicatus, and gyri hippocampi.

Increased capillary circulation of the brain may cause illusions of sensation, such as of seeing, smelling, or hearing objects not present, etc. Impressions on some of the sense-organs remain for a time, even after the cause is removed. The organs, after continued impressions, become accustomed to them unless the impression be increased, diminished, or locally changed. (See Olfactory, and Glosso-pharyngeal Nerves.) The normal stimulus on a sense-organ has its limit: if too weak there is no action; if too strong, pain.

For the sensation of touch, see tactile corpuscles and Pacinian bodies; for taste, the terminal bulb, and glosso-pharyngeal nerve; for smell, the olfactory nerve; for sight, the eye; for hearing, the ear.

CHAPTER XIX.

CRANIAL NERVES.

The **Cranial Nerves**, consisting of twelve pairs, are a part of the cerebro-spinal system, all having their origin in the gray substance of the brain, though the spinal accessory partly arises at the spinal cord. These nerves are usually enumerated, beginning at the anterior part of the cranium to the posterior (Fig. 29), in the following order:

First—Olfactory.
Second—Optic.
Third—Oculomotorius (Motor oculi).
Fourth—Trochlearis.
Fifth—Trigeminus.
Sixth—Abducens.
Seventh—Facial (Portio dura).
Eighth—Auditory (Portio mollis).
Ninth—Glosso-pharyngeal.
Tenth—Pneumogastric.
Eleventh—Spinal accessory.
Twelfth—Hypoglossal.

The cranial nerves may be classified as sensory, motor, and mixed.

The **Sensory** are the olfactory, optic, lingual or gustatory branch of the trigeminus, auditory, and a part also of the glosso-pharyngeal.

The **Motor** are the oculomotorius (motor oculi), trochlearis, abducens, facial, and hypoglossal.

The **Mixed** (sensory and motor) are the trigeminus, glosso-pharyngeal, pneumogastric, and spinal accessory.

In the medulla oblongata and pons Varolii, the nervous centre-cells giving origin to motor fibres are situated anteriorly, and on the median line; while the cells of the sensory fibres are located further inward and posteriorly, similar to the centre-cells of the motor and sensory fibres of the spinal cord, situated anteriorly and posteriorly, respectively.

FIRST PAIR OF CRANIAL NERVES—THE OLFACTORY.

The **Olfactory Nerve** carries material impulses for the sense of smell. It originates in three roots: the external, middle, and internal. The external root arises from cells of a nucleus of gray matter in the temporo-sphenoidal lobe in front of the *pes hippocampi*. The middle root arises from the corpus striatum, and from a papilla of gray matter termed *caruncula mammilaris*. The internal root arises from the *gyrus fornicatus* in the anterior lobe. The fibres of the olfactory nerve form a tract which enters a lump of gray matter termed the *olfactory bulb*, resting on the cribriform plate of the ethmoid bone. From the olfactory bulb issue about fifteen filaments, passing

Fig. 42.

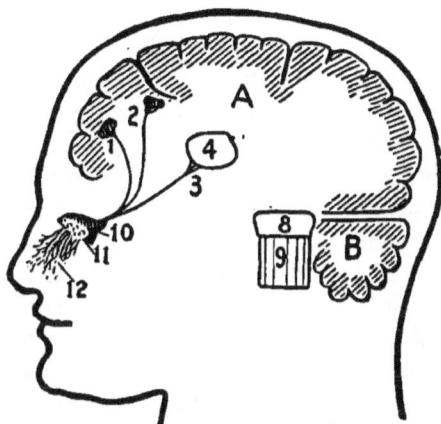

Diagram of the olfactory nerve.

A, Cerebrum. B, Cerebellum.

1, Internal root of olfactory nerve. 2, External root. 3, Middle root. 4, Corpus striatum. 8, Pons Varolii. 9, Medulla oblongata. 10, Olfactory bulb. 11, Cribriform plate of the ethmoid bone. 12, Schneiderian membrane.

through as many foramina (11) in the cribriform plate of said bone, and terminate in the olfactory or Schneiderian membrane, which is distributed through the mucous membrane of the nose. The originating centres of the olfactory nerve-fibres of both hemispheres are in communication with each other by nerve-fibrillæ.

The olfactory nerve differs from other nerves in containing soft and pulpy gray matter, its fibres being non-medullated. The size of the olfactory bulb enhances the function of smell; being comparatively large in dogs, cats, and bees, whose sense of smell is more acute. Humboldt states that the Peruvian Indians are able, in the darkness of night, to distinguish by smell whether an approaching stranger be a European, an American Indian, or a Negro. Congenital absence or disease of the olfactory bulb impairs the ability and capacity to distinguish odors. A necessary condition for smell is that the mucous membrane of the nose be moist, and that the odorous particles be in the form of gases or fine powder, and the air kept in motion by inhalation and exhalation to excite the nerve terminals. When said membrane is dry the sense of smell is either impaired or lost. Thus, in the first stage of catarrh there is little or no secretion of moisture (mucus) in the nostrils. Interruption of respiration prevents smell, while rapid inspirations and dilation of the nostrils increase and intensify it. The stimulus must come in contact with the Schneiderian membrane by inhalation. If a fluid or solid substance be applied directly to this membrane, the odor of such a substance is very slightly or not at all perceptible. The olfactory organs become accustomed to odors, and consequently it requires variation of external stimulus to produce impressions. The nasal membrane, through nasal branches of the first and second division of the fifth cranial nerve, is sensitive to cold, heat, itching, tickling, and pain. That these nerves cannot perform the function of the olfactory is proven by the fact that when smell is lost the membrane of the nose still remains sensitive to the above-mentioned sensations, all of which, however, are but modifications of the one sense of touch. The sense of smell varies in different individuals; odors pleasant to some being offensive to others. Even illusions may occasionally occur when an odor is merely imagined, but not real. The centre of smell of the external root (2) and that of taste are placed near together at the bottom of the fissure of Sylvius.

SECOND PAIR OF CRANIAL NERVES—THE OPTIC.

The **Optic Nerve** (one from each cerebral hemisphere) originates principally in the anterior portion of the corpora quadrigemina, with a few fibres from the posterior part of the optic thalamus and corpora geniculata (Fig. 43). The fibres pass inwardly forward from one side of the cerebrum towards the other, forming the two optic tracts, which meet upon the median line, where they cross each other (decussate), forming the *optic commissure*, or *chiasma* (2, Fig. 29). They then diverge, forming the optic nerves proper, which leave the cranium through the *optic foramina* and enter the orbits. At the posterior aspect of the eyeball, the nerve, piercing the sclerotic and choroid coats, terminates in a nervous expansion called the *retina*. At the decussation, each optic nerve incorporates with its own fibres a few from the origin of the opposite hemisphere; and each nerve is, consequently, in communication with both hemispheres. The optic nerve is remarkable in this, that its fibres run a longer course within than without the cranium, and furnish no branch from the origin (decussation) to the termination. Its surrounding neurilemma is derived from the dura mater. The decussation of the nerves facilitates the harmonious control of the two eyes, so that the light-rays on the retina of a single object transmit an impulse through the optic nerve of each eye to the cerebral centre of sight.

Physiologically, the optic nerve carries the impulse of stimulus for the completion of the sense of *sight*. We have said *completion*, since it requires three organs—the eye, the nerve, and the nervous centre. Normally, the rods and cones of the retina receive the stimulus of vibrations of light-ether. The impression thus produced by the luminous rays upon the retina causes the sensation of sight as long as the retina is connected with the nervous centre by the optic nerve. Section of the optic nerve is not painful, but

Fig. 43

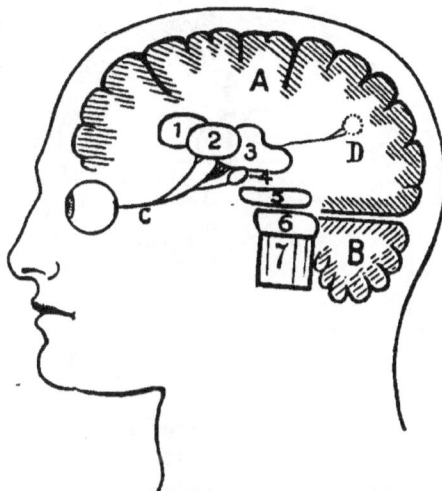

Diagram of the Optic Nerve.

A, Middle lobe of the cerebrum. B, Cerebellum. C, Optic nerve. D, Conscious sensation centre, or cerebral area of sight, in the posterior cerebral lobe.

1, Corpus striatum. 2, Optic thalamus. 3, Corpora quadrigemina. 4, Corpora geniculata. 5, Crus cerebri. 6, Pons Varolii. 7, Medulla oblongata.

produces blindness immediately in the eye affected, since impressions received by the retina are no longer transmitted to the cerebral centres. The partial blindness of both eyes is sometimes observed in hemiplegia, in which probably the transmission of light becomes defective through both the crossed and direct fibres of the optic tracts. A stimulus applied to one eye may cause contraction of the pupils of both eyes, because sensitive impressions on the retina, conveyed only by one optic nerve to the nervous centre of sight in the cerebrum, are transformed into motor impulses, and by reflex from the centre of sight sent to the iris simultaneously through both of the oculomotorius nerves. (*Vide* Eye and Sight.)

THIRD PAIR OF CRANIAL NERVES—THE OCULOMOTORIUS.

The **Oculomotorius**, or **Motor Oculi** nerve, carries impulses of voluntary motion to the levator palpebræ, to the superior, internal, and inferior recti, and inferior oblique muscles of the eye; *i. e.*, it supplies four out of the six muscles that move the eyeball. Its fibres originate from the valve of *Vieussens*, situated between the third and the fourth ventricles, and some fibres are traceable to the corpora quadrigemina, crus cerebri, and the floor of the fourth ventricle. The fibres aggregate and pass to the orbit through the sphenoidal fissure, between the two heads of the external rectus muscle. The physiological properties of the oculomotorius nerve are to carry exclusively motor impulses to the upper eyelid, and for the vertical, lateral, and rotary movements of the eyeball, but contractile for the iris and ciliary muscles. In paralysis of this nerve there is drooping of the eyelid (ptosis), external strabismus, and dilatation of the pupil. (*Vide* Eye and Sight.)

Fig. 44.

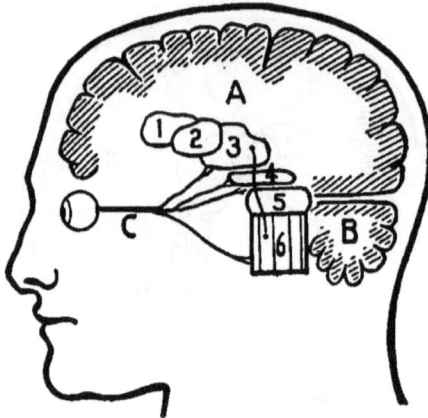

Diagram of the oculomotorius, or motor oculi, nerve.

A, Cerebrum.   B, Cerebellum.   C, Oculomotorius nerve.

1, Corpus striatum.   2, Optic thalamus. 3, Corpus quadrigeminus. 4, Crus cerebri. 5, Pons Varolii. 6, Medulla oblongata. The line with end dots, from corpus quadrigeminus to the medulla oblongata, indicates the intimate connection of these two centres with each other, in regard to the function of this nerve.

FOURTH PAIR OF CRANIAL NERVES—THE TROCHLEARIS.

The **Trochlearis**, or **Patheticus** (one from each hemisphere), is the smallest cranial nerve (Fig. 45). Its fibres originate at the lateral walls of the valve of *Vieussens*, also at the posterior part of the crus cerebri just beneath the corpora quadrigemina, and some fibrillæ from the upper anterior part of the fourth ventricle. Passing then to the floor of the Sylvian aqueduct, and through the sphenoidal fissure into the orbit, it is distributed exclusively to the superior oblique muscle of the eyeball. It possesses the physiolog-

ical property of transmitting the material impulses for involuntary motion to said muscle, act-
ing in harmony with the oculomotorius nerve to pre-serve the horizontal plane of the eyeball; but in co-ordinated movements, the trochlearis carries the ma-terial impulses which are under the control of the will. Its paralysis inca-pacitates the rotation of the eyeball and double vision *(diplopia)* ensues. Le-sion at the origin of its fibres at one hemisphere af-fects the eyeball on the op-posite side, since the great-er number of fibres of this nerve decussate near the origin.

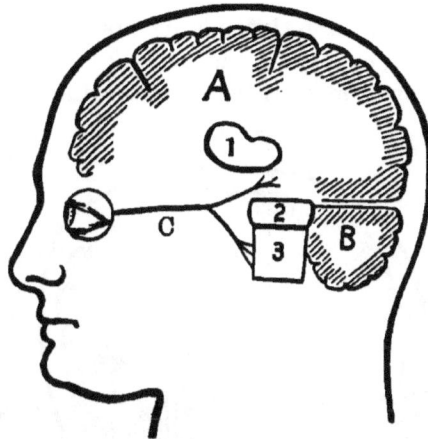

Fig. 45.

Diagram of the trochlear nerve.

A, Cerebrum. B, Cerebellum. C, Troch-lear nerve.

1, Corpus quadrigeminus. 2, Pons Varolii. 3, Medulla oblongata.

FIFTH PAIR OF CRANIAL NERVES—THE TRIGEMINUS.

The fifth nerve (one from each hemisphere) is termed **Trigem-inus** *(threefold)*. It originates from two roots of fibres: the first, a large posterior root (sensory) of about seventy to one hundred nerve filaments springing from a nucleus of cells on the floor of the fourth ventricle and on the posterior horn of gray matter of the spinal cord, as far down only as the second vertebra, also from the lateral portion of the pons Varolii; the second, a small anterior root (motor) of about twenty nerve filaments, arising from a nu-cleus of cells at the lateral angle of the fourth ventricle and pyr-amidal body.

The trigeminus nerve includes more nerve-fibres than any other cranial nerve, and passes to the **Gasserian ganglion** *(semilu-nar)* (D, Fig. 46), lodged in a depression near the apex of the pet-rous portion of the temporal bone. Its fibres carry material im-pulses for three different functions: special sense (taste), common sensory—*i. e.*, a function involving diverse external senses, and mo-tory. It is the great sensitive nerve of the skin of the head, face, and nasal mucous membrane, also the motor nerve for the muscles of mastication. The Gasserian ganglion gives off three nerves: the ophthalmic (E), superior maxillary (H), and inferior maxillary (K).

The **Ophthalmic** is its first division, and carries sensory impulses, supplying the eyeball, lachrymal gland, mucous lining of the eye and nose, the integument and muscles of the eyebrow, nose,

Fig. 46.

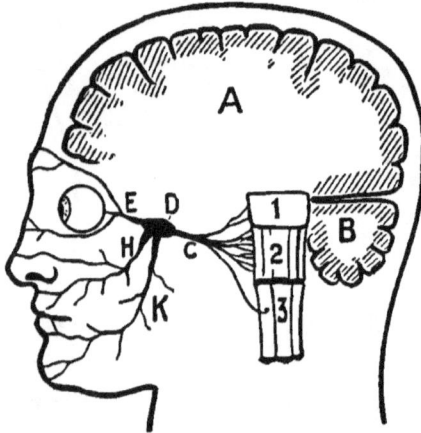

Diagram of the trigeminus nerve.

A, Cerebrum. B, Cerebellum. C, Fifth, or trigeminus nerve. D, Gasserian ganglion. E, Ophthalmic nerve. H, Superior maxillary nerve. K, Inferior maxillary nerve. 1, Pons Varolii. 2, Medulla oblongata. 3, Cervical portion of the spinal cord.

and forehead. It consists of fibres from the origin of the fifth pair and some additional from the Gasserian (semilunar) ganglion. The ophthalmic divides into three branches: lachrymal, frontal, and nasal, which pass through the sphenoidal fissure into the orbit. If its nasal branch be injured or irritated, the mucous membrane of the nasal passage swells and becomes obstructed by mucus, which assumes a fungus consistency, and is liable to bleed at the slighest touch. The sense of smell is then marred, as in influenza, while the olfactory nerves are inactive on account of the alteration of the mucous membrane and its secretions.

The **Superior Maxillary** is the second division of the fifth nerve. This also carries sensory impulses. It contains fibres from the origin of the fifth, and some additional from the Gasserian (semilunar) ganglion, then passes through the foramen rotundum, and supplies the teeth, gums, mucous membrane of the upper jaw, lower part of the nose, the integument of the lower eyelids, the muscles of the nose, temple, and side of forehead, the skin, and mucous membrane of the upper lip.

The **Inferior Maxillary** is the third division of the fifth nerve. It is a mixed nerve—i. e., both sensory and motor. It also contains fibres from the origin of the fifth, and some additional from the Gasserion ganglion, and passes out of the skull through the *foramen ovale* in the sphenoid bone, supplying the teeth and gums of the lower jaw, integument of the temple and ear, lower part of the face and lower lip, the muscles of mastication, and the tongue with a nerve for its special sense of taste.

The three branches just described constitute principally one of the fifth pair of cranial nerves, their physiological property being principally to convey sensory impulses. The parts supplied by this nerve are the seat of all neuralgic affections about the head and face. The lingual branch of the inferior maxillary is not only necessary to the sense of taste, but it also enables us to realize when food is uniformly reduced by mastication, and whether some is left about the mucous membrane or teeth. With a tongue dry or coated, as in febrile conditions, taste is marred, showing that it depends upon the special sensitiveness of the papillæ of the lingual nerve on the tongue, as well as upon its moisture—*i. e.*, on all the physiological conditions required for the integrity of the mucous membrane. Its muscular branch supplies the temporal and masseter muscles which bring the teeth of the lower jaw most powerfully in contact with the upper. Section of this nerve paralyzes the muscles of mastication; but if paralyzed on one side only, mastication may continue by lateral motion of the jaw.

The fifth nerve may also affect sight. Impressions of light are transmitted exclusively by the optic nerve, but the anterior part of the eyeball, iris, and cornea are connected by sensory fibrillæ with the ophthalmic ganglion. If the trigeminus nerve be divided or injured within the cranium, or the Gasserian ganglion, it causes a change of nutrition in the cornea by vascular congestion, ulceration, and destruction of the eye; indicating that the impulses influencing nutrition of the eyeball are not by its own proper nerve-fibres of the trigeminus only, but also by additional fibrillæ carrying impulses from the Gasserian ganglion. The cerebral areas, or centres of taste and smell, are placed near together at the bottom of the fissure of Sylvius.

SIXTH PAIR OF CRANIAL NERVES—THE ABDUCENS.

The **Abducens Nerve** (one from each hemisphere), originates by fibrillæ from the lower posterior part of the pons Varolii and upper part of the anterior pyramid, or corpus pyramidale; also from the fasciculus teres on the floor of the fourth ventricle. It carries motor impulses to the external rectus muscle, effecting abduc-

12

tion of the eyeball. The nerve-fibrillæ of the upper origin, with those of the opposite partner, bend over the pons and join those of the deeper origin. The fibres then pass through the sphenoidal fissure, forming the abducens nerve, which supplies the external rectus muscle of the eye. The fibres bending over the pons may be easily pressed upon by any abnormality of the tentorium cerebelli; hence, both nerves are very liable to partial paralysis, causing strabismus (squinting), or *diplopia* (double-sight—*i. e.*, seeing a single object as two). Squinting may also be caused by partial opacity of the cornea, or by injuries, shortness, or contraction of the external rectus muscle.

Fig. 47.

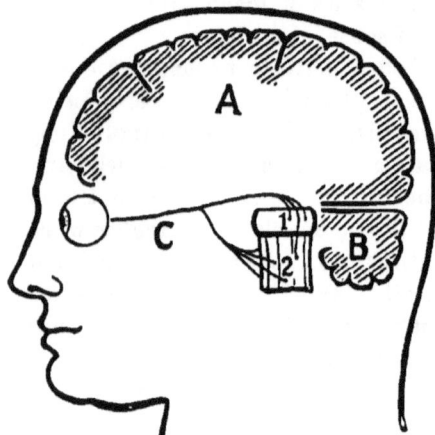

Diagram of the abducens nerve.

A, Cerebrum. B, Cerebellum. C, Abducens nerve.

1, Pons Varolii. 2, Medulla oblongata.

SEVENTH PAIR OF CRANIAL NERVES—THE FACIAL.

The **Facial Nerve** (*portio dura*, Willis), one from each hemisphere, originates by fibres from the lateral tract of the medulla oblongata, between the olivary and restiform bodies, and partly from the fasciculus teres on the floor of the fourth ventricle. It passes in company with the auditory nerve (8th) into and through the internal auditory meatus, where the facial continues through the petrous portion of the temporal bone, coming out at the stylo-mastoid foramen, and is distributed to the side of the face in a fan-shaped manner. It communicates with the auditory, glosso-pharyngeal, and pneumogastric nerves; with the spheno-palatine (Meckel's), and otic ganglia; carotid and cervical plexuses; parotid gland, and terminals of the fifth nerve in the integument of the face. The *chorda tympani* nerve branches off from the facial nerve just before it makes its exit from the stylo-mastoid foramen; it then passes to the lingual muscle.

The facial nerve carries motor impulses to the muscles of the face, to those of the external and middle ear, palate, and tongue.

Its impulses on the muscles of the face change the features in their various movements to correspond with the varying phases of mental or emotional activity. A lesion or pressure on this nerve paralyzes all the superficial muscles of the face, thereby causing loss of expression on the paralyzed side.

The nostrils in man are more or less rigid, and nearly inactive in their ordinary condition. They expand when the air is forcibly inspired, to assist in the sense of smell, or whenever the breathing is hurried or laborious through increased muscular exertion, or by any accidental obstruction of the air-passages. In case, however, of facial paralysis, the nostrils become compressed by inspiration, while during expiration they are forced outward. In this case their natural movements in respiration are reversed. Such paralysis may be caused by disease of the facial nerve itself, disease at its origin in the brain, or by disease of the bony canal at the petrous portion of the temporal bone. If the lesion is located in the peripheral branches of the nerve, certain facial muscles may be affected to the exclusion of others; for instance, the muscles about the lips may be paralyzed without any perceptible loss of motion in the parts above; but if located on the central trunk of the nerve, or if it involves the whole of its origin, then the consequences extend uniformly over one side of the face, forming a complete unilateral facial paralysis (Bell's). As the muscles on the paralyzed side are no longer antagonistic to those of the other, the face is drawn over to the healthy side, and may be accompanied by disturbances of mastication, deglutition, speaking, and taste. The secretion of saliva may also be interfered with. The orbicularis palpebrarum likewise becomes powerless, the eye remains open, the angle of the mouth is lower, and during mastication some

Fig. 48.

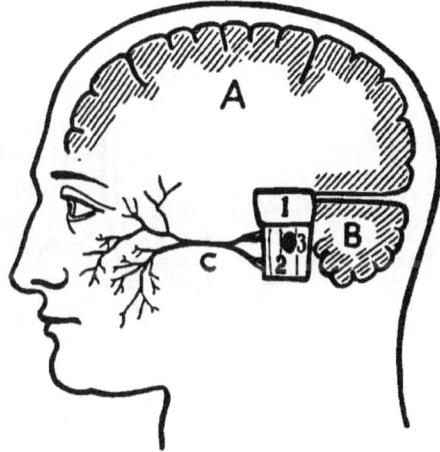

Diagram of the facial nerve.
A, Cerebrum. B, Cerebellum. C, Facial nerve.
1, Pons Varolii. 2, Medulla oblongata. 3, Olivary body.

food remains between the cheek and gums. An abnormality of taste on the same side as the facial paralysis is a symptom that fixes the location of the lesion of the facial nerve, at some point inside the stylo-mastoid foramen.

The **Auditory Nerve,** or *nervus acusticus*, is soft, and hence is termed *portio mollis* of the 7th (Willis). It arises from three roots: one large anterior root and a small posterior one, each at a median and a lateral nucleus. These four nuclei (*tubercula acustica*) are situated in the floor of the fourth ventricle in connection with the posterior pyramid and restiform body. The fibres of these two roots carry the impulses from the cochlea and vestibule of the internal ear for hearing. The third root arises at the superior vermiform process of the cerebellum (Fig. 49), and carries the impulses from the semicircular canals of the ear labyrinth for maintaining the equilibrium of the head and of the body. All the fibres of the three roots combine and enter the internal auditory meatus, where this nerve receives a few fibres from the facial nerve (7th). It passes onward to the internal ear, and is distributed to the vestibule, cochlea, and semicircular canals.

Fig. 49.

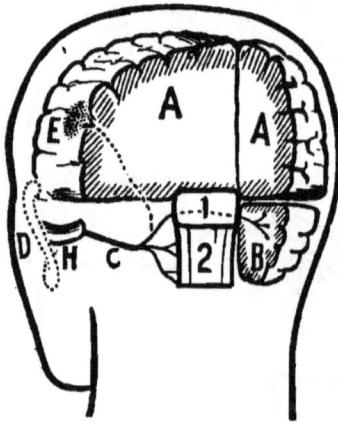

Diagram of Auditory Nerve.

AA, The occipital bone partly removed, exhibiting the posterior portion of the brain. B, Cerebellum. C, Auditory nerve. D, External ear. H, Internal auditory meatus. E, Sensation centre of hearing. 1, Pons Varolii. 2, Medulla oblongata.

The auditory nerve carries impulses for the sense of hearing, and governs the equilibrium of the body, as it is intimately connected with the auditory area in the cerebrum, also with the centres of the sense of position of the head and body, and with the sense of direction of space. Stimulation of the auditory nerve in any part of its course, or at its termination in the vestibule or cochlea, produces the sensation of sound. Stimulation at its termination in the semicircular canals determines the proper position of the head and body; while any irritation to said canals disturbs the equilibrium, more or less, of the head or body, or of both; also disturbing the proper direction. (*Vide* Semicircular Canals in chapter on Ear and Hearing.)

NINTH PAIR CRANIAL NERVES—THE GLOSSO-PHARYNGEAL.

The **Glosso-pharyngeal Nerve,** one from each hemisphere, contains sensory and motor fibres, some of which arise behind the olivary body of the medulla oblongata, carrying impulses for the sense of taste and for deglutition. Some fibres arise from a nucleus of cells at the lower half of the floor of the fourth ventricle, carrying the sensory impulses from the mucous membrane of the pharynx, fauces, and tonsils. After leaving the centres of origin, the fibres join and the nerve proper is formed, receiving its neurilemma from the dura mater, just when passing from the cranium in company with the pneumogastric and spinal accessory nerves through the jugular foramen. At this foramen the nerve passes through the jugular ganglion, and at the petrous portion of the temporal bone proceeds through the petrous, or *Andersh's,* ganglion.

Fig. 50.

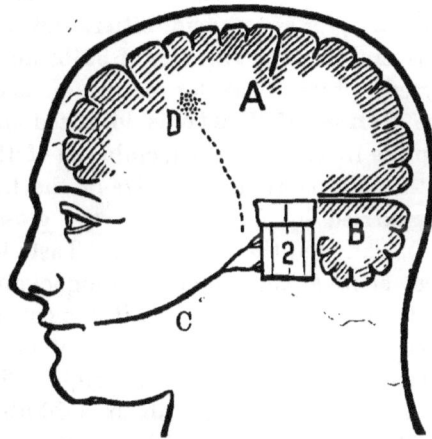

Diagram of the glosso-pharyngeal nerve.

A, Cerebrum. B, Cerebellum. C, Glosso-pharyngeal nerve. D, Area, or centre, of taste.
1, Pons Varolii. 2, Medulla oblongata.

It sends off one branch to the tongue, and another to the pharynx, tonsils, and soft palate. Its function is to transmit material impulses for two different, but associated, phenomena: 1, the sense of taste; 2, the impulses of reflex action for deglutition. The latter at first is voluntary; but as soon as food has passed the epiglottis, deglutition is performed involuntarily by reflex action, even during unconsciousness. The stimulus for deglutition is the food or liquid on the surface of the mucous membrane of the fauces and pharynx. The impulse is conveyed by the glosso-pharyngeal nerve inward to the medulla oblongata, which by reflex action forwards a motor impulse, passing with fibres of the pneumogastric to the pharynx. Natural stimulants, such as proper food or liquid in contact with the mucous membrane of the pharynx, excite deglutition; while unnatural substances, like a finger or a feather, produce vomiting. Nauseous or irritating substances, too, cause more or less abnormal reflex actions to the muscular fibres in the walls of the throat, œsophagus, stomach, and diaphragm, resulting in vomiting.

The **Tongue** is supplied with three nerves—the lingual (gustatory), *sensory*; hypoglossal, *motor*; and glosso-pharyngeal, *sensory* and *motor*. The sense of touch on the tongue indicates that this organ contains, at least, four varieties of terminal nerve-bulbs of taste; *i. e.*, for sweet, acid, bitter, and saline. The function of one or of several of these varieties of bulbs may become diminished or paralyzed, while those for saline or acid tastes, are not affected. The **Sense of Taste** is localized in the terminal bulbs of the papillæ in the mucous membrane of the tongue, and slightly also at the soft palate and pillars of the fauces. Its sensory impulses are carried by two nerves — the glosso-pharyngeal and the lingual branch of the fifth nerve. Taste for sweet and acid matters is most acute at the tip of the tongue, and less so on the back part; while for bitter and saline, it is most acute at the back part, and least at the tip. All sensations of taste are more acute on the left edge of the tongue than on the right. The terminal bulbs (Fig. 36) are minute oval bodies about 1-300th of an inch in length, and 1-800th of an inch in width, and where they are the most numerous there the sense of taste is the most acute. An increased sense of taste is called *hypergeusia*; a diminished is called *hypogeusia*; and the loss of taste is termed *ageusia*. To excite a sensation of taste, the stimulating substance must be soluble in the fluid of the mouth. An insoluble substance in contact with the tongue causes the sensation of touch or temperature only, but excites no taste. Smell is closely associated with taste, for, if a substance is tasted, its aroma (smell) is perceptible at the same time. If the nose be closed during the act of swallowing or tasting, the sense of taste is greatly diminished. It also requires that the substance tasted be in a state of solution, as a solid substance causes the impression only of a foreign body. Substances with an unpleasant taste, when placed on the back part (root) of the tongue, often induce vomiting, but have little or no nauseous effect when placed on the forward part. The glossal branch of the glosso-pharyngeal nerve is distributed to the mucous membrane and papillæ of the root of the tongue, and carries the impulse of sensation of taste of that part. The hypoglossal nerve supplies the muscular portion of the tongue; while the lingual (gustatory) branch of the fifth nerve carries the impulse of sensation from the taste bulbs in the mucous membrane of the anterior two-thirds of the tongue. Substances in solution penetrate to the papillæ of the tongue by endosmosis and by attraction of the cells in the papillæ. With a coated tongue, therefore, taste is more or less vitiated. An impression on the terminal bulbs of the papillæ is retained for some time, so that in tasting different flavors repeatedly in quick succession no difference can be distinguished. By tasting a highly flavored medicine nauseous drugs may be swallowed pleasantly. (See Terminal Bulb.)

TENTH PAIR OF CRANIAL NERVES—THE PNEUMOGASTRIC.

The **Pneumogastric (Vagus) Nerve** (Fig. 51), one from each hemisphere, is composed of sensory and motor fibres. The fibres of its origin proper, are sensory only, and arise from the posterior part of the floor of the fourth ventricle. Its motor fibres are derived from other sources, since it receives nerve-fibres from the spinal accessory, glosso-pharyngeal, and hypoglossal. It gives off ten branches; viz., auricular, pharyngeal, laryngeal (superior, and recurrent), cardiac (cervical, and thoracic), pulmonary (anterior, and posterior), œsophageal, and gastric. The fibres, after leaving the centres of origin, join, and the nerve proper is formed, receiving its neurilemma from the dura mater just when passing from the cranium in company with the glosso-pharyngeal and spinal accessory nerves through the jugular foramen. It supplies the pharynx, larynx, organs of the voice, trachea, œsophagus, lungs, heart, thorax, stomach, and greater part of the intestines. This nerve is the channel through which the sensory impulses for respiratory and circulatory movements are conveyed to the medulla oblongata, which sends forth reflex impulses to the lungs, respiratory muscles, and to the heart. It also wields an important influence in digestion, as it supplies the walls of the stomach with sensory nerve-fibres. The reflex impulses carried by its motor fibres are derived not only from the respiratory and circulatory centres, but also from other centres, the fibres of which run within the pneumogastric nerve, such as those contributing from the spinal accessory, glosso-pharyngeal, hypoglossal, and others already mentioned. The reflex impulses carried by its motor fibres call forth not only normal activities of already stated organs, but also (if irritated) a variety of abnormal reflex phenomena, such as coughing, and vomiting. (See Respiration, Circulation, Heart, Arteries, Stomach, Saliva, and Vcice.)

Fig. 51.

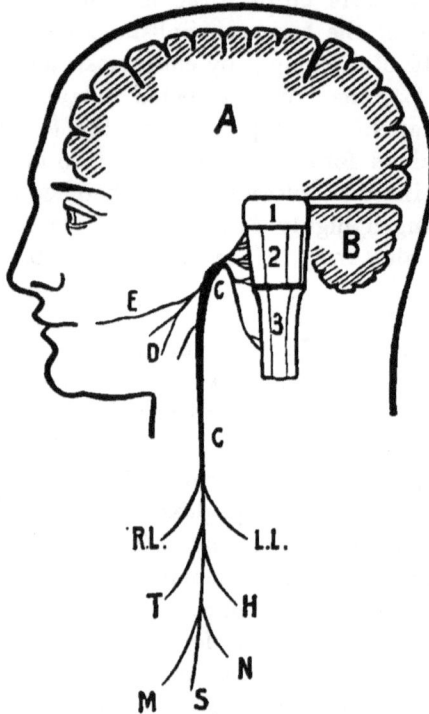

Diagram of the pneumogastric nerve.

A, Cerebrum. B, Cerebellum. C, Pneumogastric nerve. D, Laryngeal branches. E, Pharyngeal branch. R. L, Pulmonary branch to right lung. L. L, Pulmonary branch to left lung. T, Œsophageal branch. H, Cardiac branch. N, Gastric branch joining the cœliac and splenic plexuses. S, Gastric branch to anterior surface and curvatures of the stomach. M, Branch joining the hepatic plexus.

1, Pons Varolii. 2, Medulla oblongata. 3, Cervical portion of the spinal cord.

ELEVENTH PAIR OF CRANIAL NERVES—THE SPINAL ACCESSORY.

The **Spinal Accessory Nerve** (Fig. 52), one from each half of the medulla oblongata and spinal cord, arises from two roots, one from cells near the vagus nucleus on the floor of the fourth ventricle in the medulla oblongata, the other from cells in the anterior and posterior horns of gray matter in the spinal cord between the 4th and 6th cervical vertebræ (3, Fig. 52). From this latter origin emerge a series of separate rootlets that collectively pass upward through the foramen magnum, there joining the fibres from the medulla, then incorporating with the pneumogastric (*vagus*)

Fig. 52.

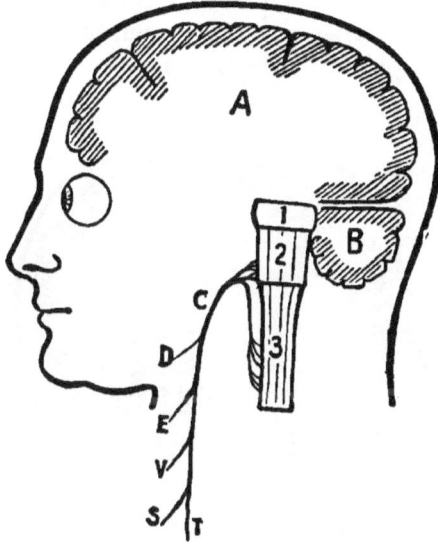

Diagram of the spinal accessory nerve.

A, Cerebrum. B, Cerebellum. C, Spinal accessory nerve, where it runs in the same sheath with the pneumogastric nerve. D, Branch communicating with the pharyngeal branch of the vagus. E, Branch communicating with the superior laryngeal branch of the vagus. V, Branch supplying the sterno-mastoid muscle. S and T, Branches supplying the trapezius muscle.

1, Pons Varolii. 2, Medulla oblongata. 3, Cervical portion of the spinal cord.

nerve, with which they pass from the cranium through the jugular foramen. It supplies, either directly or by anastomosis, motor influence to the pharynx and larynx, and the sterno-mastoid and trapezius muscles. Some of its fibres are connected with the cardiac plexus. It controls, also, the vocal organs. Section or sufficient injury to it causes loss of power to produce vocal sounds. Other movements, such as the peristaltic in the larynx for deglutition, are not interfered with. Section of the pneumogastric, however, paralyzes the glottis, the organs of respiration, phonation, and deglutition.

The **Hypoglossal Nerve** (one from each hemisphere) arises from some fibres in a groove between the pyramidal and olivary body, and some from a nucleus of cells at the anterior and lowest portion of the fourth ventricle. The fibres join, and the nerve proper is formed by receiving its neurilemma from the dura mater just when passing from the cranium through the anterior condyloid foramen. The fibres all arise anteriorly, and, therefore, physiologically carry *motor* impulses to the tongue and most of the muscles

Fig. 53.

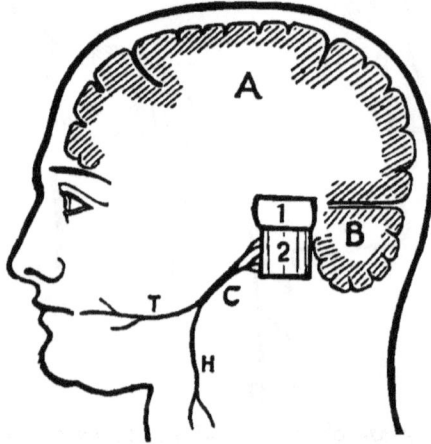

Diagram of the hypoglossal nerve.

A, Cerebrum. B, Cerebellum. C, Hypoglossal nerve. T, Its muscular branches to the tongue. H, Its descendens noni branch.
1, Pons Varolii. 2, Medulla oblongata.

arising from the hyoid bone. Its fibres of the lower origin decussate at the median line; therefore, cases of facial paralysis may be accompanied by paralysis of the tongue on the same side with that of the face, being the side opposite the lesion. One of the genio-hyo-glossal muscles in this case loses its power, while the muscle opposite remains active; so that in protrusion of the tongue it is drawn to one side. The motor influence of this nerve is essential in mastication, deglutition, and in articulation of speech. Section or injury to the nerve either stops or interferes with the reflex actions for these functions.

CHAPTER XX.

THE EYE AND THE SENSE OF SIGHT.

The **Eyeball** (*bulbus oculi*) (Fig. 54) is the organ of *sight*; and, considered as an optical instrument, may be said to resemble a camera obscura. It is held in position in the orbit by the surrounding muscles, eyelids, and optic nerve.

The **Sclerotica** or **Sclerotic Coat** (Fig. 54) is a tough, resisting membrane, which surrounds the eyeball externally. Its anterior segment is termed the **Cornea,** and is composed of five layers of horny, fibrous, unyielding tissue, and is perfectly transparent. The cornea has no blood-vessels, but is supplied by nerve-fibres from the ciliary nerves. It differs in this respect from the other portions of the **sclerotica**, which consists of white and elastic fibrous tissue, opaque, very hard and dense, and forms an unyielding membrane, impervious to light. The sclerotica receives the insertions of the recti and oblique muscles of the eyeball, and is well supplied with ciliary vessels and nerves. The cornea may become altered in structure through pressure or tension, in which case it presents an opaque, milky appearance, like the rest of the sclerotic coat.

The **Choroid** coat of the eye (Fig. 54) is a highly vascular, opaque, blackish brown pigment membrane lying between the sclerotica and retina. Its dark color and density prevent dispersion of the rays of light within the eye, thereby favoring the absorption by the retina of all the light that enters the eye.

The **Hyaloid Membrane** is a thin, soft, gelatinous, structureless membrane surrounding the vitreous humor, into which it sends numerous transparent, clear processes for the support of the humor. These processes serve to equalize the pressure of the surrounding membranes. (*Vide* Vitreous Body.)

The **Vitreous Body,** or **Humor**, is a soft, gelatinous, transparent, glossy structure, filling about four-fifths of the eye-globe posteriorly, where it is surrounded by the retina. It is composed of water, mucin, salts, and albumen. It contains no blood-vessels, its nutrition being imbibed from vessels of the retina and ciliary pro-

cesses. Possibly, the amœboid and multipolar cells contained in
the vitreous body are more or less stimulated to activity by the different color-vibrations, whereby a difference of stimuli on the retina may be produced.

The **Aqueous Humor** is a limpid liquid, consisting of nearly
pure water, with traces of albumen and chloride of sodium. It is
enclosed in a cavity between the cornea and lens (Fig. 54). This
cavity is divided by the iris into two chambers—an anterior and a
posterior. The **Anterior Chamber** is bounded by the cornea,
iris, ciliary muscle, and ligament; and is filled with aqueous humor. The **Posterior Chamber** is bounded by the iris, lens, and
suspensory ligament; it also is filled with aqueous humor.

Fig. 54.

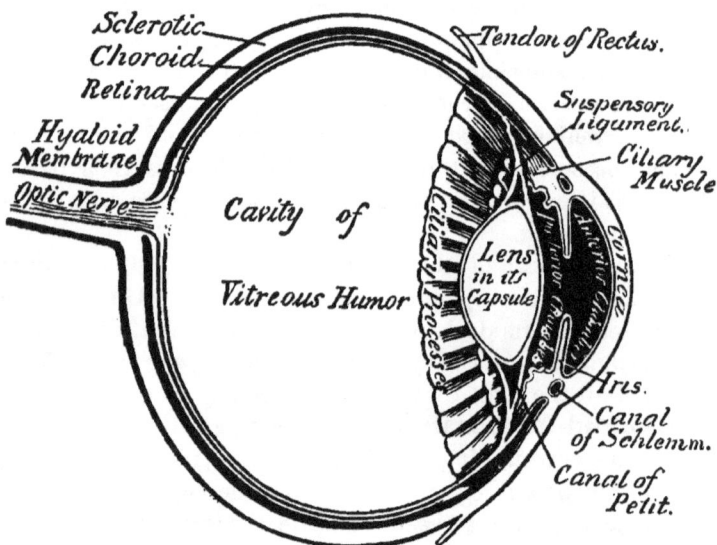

Diagram of the eye.

The **Ciliary Muscle** forms a circular band, about one-eighth
of an inch broad, lying on the fore part of the choroid coat, thickest in front and tapering behind in the part attached to the choroid, just in front of the retina. When this muscle contracts it
compresses the vitreous humor; or, possibly, by pulling on the ciliary processes, at the same time relaxing the suspensory ligament, it
assists the lens to protrude forward (increases its convexity), thus
facilitating accommodation of sight for different distances. Its arteries and nerves are derived from the ciliary arteries and nerves.

The **Ciliary Ligament** consists of circular, elastic fibres, about one-fourth of an inch thick, anterior to the ciliary muscle, and containing the *canal of Schlemm*. It surrounds the iris, and connects together the cornea, sclerotica, iris, and outer fore part of the choroid.

The **Suspensory Ligament** of the lens is a transparent, tough, elastic, fibrous membrane between the vitreous humor and front portion of the ciliary processes, connecting with the border of the lens near its circumference, thereby constituting by its anterior layer a part of the boundary of the posterior chamber. Its internal layer is attached to the ciliary processes. It is also connected with the anterior portion of the retina, assisting in retaining the lens in position.

The **Canal of Petit** is a triangular passage, about one-tenth of an inch in size, around the border of the lens, lying within the two layers of the suspensory ligament.

Pupil, action of: *Vide* Iris.

The **Iris** (rainbow-like, on account of its many colors) is a thin, circular, contractile, muscular curtain, suspended behind the cornea, but in front of the lens (Fig. 54). Its color constitutes the color of the eye (so-called). The orifice in the center is the *pupil*—"the window of the eye"—whose function is the regulation (by involuntary enlargement and contraction) of the quantity of light admitted to the interior chamber of the eye. This regulative function prevents strong light from dazing the vision, or injuring the retina. A few strong rays affect the retina more severely than many weak ones. The iris divides the cavity between the cornea and lens into two chambers (an anterior and a posterior), which freely communicate in the center through the pupil (Fig. 54). When rays of light focus on the retina the iris contracts and the pupil becomes small, through reflex action; the afferent nerve being the optic, the efferent, the oculomotorius through the ciliary fibres from the ophthalmic ganglion. The movements of the iris are involuntary. The impression of light is conveyed by the optic nerve to the anterior tubercle of the corpora quadrigemina, which transfers the impulse to the valve of Vieussens, crus cerebri, and medulla oblongata (Fig. 43, and Fig. 44). There the oculomotorius nerve (third) originates, through which is transmitted a reflex impulse passing to the ophthalmic ganglion, thence through the ciliary nerves to the iris, for contraction. The contraction is therefore produced through reflex action. The destruction of the corpora

quadrigemina abolishes the reflex centre, and the iris (pupil)
remains dilated, whatever be the intensity of the light. The iris
contracts and the pupil becomes small, not only through the influence of intensified light, but also from any condition producing
congestion of the blood vessels of the eyeball; from accommodation for near objects; and from the use of nicotine, opium, or phy-
sostigmine. Passing suddenly from light to darkness obscures
sight, until a sufficient dilatation of the iris and consequent enlargement of the pupil has taken place; but passing from darkness
to bright light dazes the eye from inability of the iris to contract
with sufficient rapidity. Strong light on one eye causes contraction of the pupils of both eyes, the contraction being due to an impulse through both oculomotorius nerves from within outward.
A brilliant light may so contract the iris that its traction on the
ciliary and suspensory ligaments will injure the retina. The reflex
impulses by which the iris (pupil) contracts (see corpora quadrigemina) or expands (see cilio-spinal centre) under varying degrees
of light pass through the ophthalmic ganglion. The iris, therefore,
acts sluggishly because the intervening ganglion belongs to the
sympathetic system. Section of the oculomotorius nerve produces
permanent dilatation of the iris. The centre for its normal dilatation lies at the lower cervical portion of the spinal cord, from
which nerve-fibrillæ issue to the cervical sympathetic ganglion,
and then pass to the ophthalmic ganglion. Section of these fibrillæ
near the spine causes permanent contraction of the iris. It is
stated above that any condition producing congestion of the blood-
vessels causes the iris to contract. Now, any cause relieving the
blood-vessels produces dilatation; such as diminished light, vision
for a distant object, mental depression, or such drugs as atropia
and hyoscyamine.

Nerves.—The circular muscle-fibres of the iris are supplied by
branches of the short ciliary nerve from the ophthalmic ganglion.
The fibres are derived primarily from the oculomotorius nerve and
from the cervical sympathetic ganglion. The elongated or radiating muscular-fibres of the iris are supplied by two or three long
ciliary nerve-branches from the nasal nerve, running with the short
ciliary from the ophthalmic ganglion to the iris. Under normal
conditions the sympathetic impulses are much more powerful, and
may overcome contraction. Some stimulants, such as bright light,
opium, calabar bean, tobacco, chloroform, and alcohol (first stage),
stimulate the contracting centre and paralyze the dilator; while
others, such as darkness, fear, horror, or such drugs as atropia, hyos-

cyamine, chloroform, and alcohol (latter stage), stimulate the dilating centre, and paralyze the contractor.

The **Lens** is a crystalline, double-convex, transparent body. Regarding its position in the eye, it is about a third of an inch in transverse and about one-fourth in antero-posterior diameter (Fig. 54). It weighs three to four grains. It is composed of layers of fibres, the density of which increases from without inward, the central or nucleus portion being the most dense. Its substance consists of about 70 per cent water, 25 per cent crystalline solid matter, and about 5 per cent of albumen, fat, salts, and cholesterin. It refracts the rays of light to a focus on the retina. (*Vide* Accommodation of the Eye.)

The **Ciliary Processes**, numbering about sixty to eighty, large and small, lie side by side, and are attached to folds running backward from the posterior margin of the suspensory ligament to the anterior part of the retina (Fig. 54).

The **Retina** is a soft, delicate, semi-transparent membraneous expansion of the optic nerve. It is situated between the hyaloid

Fig. 55.

Diagram of the layers of the retina.

1, Retina. 2, Pigmentary layer. 3 and 4, Jacob's membrane of rods and cones (3, the rods; 4, the cones). 5, Membrana limitans externa. 6, Outer nuclear layer. 7, Outer molecular layer. 8, Inner nuclear layer. 9, Inner molecular layer. 10, Vesicular layer. 11, Layer of optic nerve-fibres. 12, Membrana limitans interna, with blood-vessels and optic nerve-fibres.

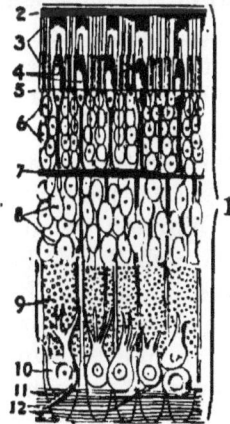

membrane and choroid coat, extending from the entrance of the optic nerve outward and forward to the ciliary processes (Fig. 54). Its greatest thickness, at the entrance of the optic nerve, is only about 1-120th of an inch. As it proceeds forward it diminishes to 1-200th of an inch. The sensitive elements of the retina are in the *macula lutea* (*limbus luteus*), termed *Jacob's membrane*. This is the layer of rods and cones on which luminous rays have their peculiar effect. The macula lutea is a roundish yellow pigment spot, about 1-20th of an inch in diameter, on the retina directly on the axis of vision, where the layers of the retina are much thinned. (*Vide* Light and Sight.)

The **Optic Nerve** (Fig. 43, and Fig. 54) contains only sensory fibres, transmitting the normal impulses received from the retina to the visual centre of the brain, which is required, as we have seen in the chapter on sensation, for the production of the external sensation of light. If the optic nerve in any part of its course is artificially stimulated, either mechanically or electrically, a sensation of light is also produced, as such stimulation, too, is transmitted to the rods and cones of the retina. Luminous impressions in the eye, as sparks, for instance, may be caused by pressure externally on the eyeball when in the dark, or by fatigue of the retina from too strong light or too continued gaze, such as in reading, and afterwards closing the eyes. (See Optic Nerve.)

The **Movements** of the **Eyeball** are produced by six muscles arising from the contour of the optic foramen and vicinity, and attached to the sclerotic coat. Four muscles are straight, and termed external, internal, superior, and inferior recti, moving the eye outward, inward, upward, and downward. Whenever the external or the internal rectus muscle permanently contracts, or the opposing muscle is paralyzed, *strabismus* is the result. The impulses through the oculomotorius (third) nerve principally move the muscles of the eyeball. The superior oblique muscle, however, is actuated through the trochlearis (fourth), the external rectus through the abducens (sixth).

Nerves and **Arteries.**—The sensory impulses of sensations of the eyeball and lid are carried through the ophthalmic branch of the fifth nerve. By and through the ophthalmic and ciliary ganglia, the sensory branches of the fifth, the motor branches of the third, and the sympathetic filaments of said ganglia, are united. Among the various parts that make up the eye the terminal cones and rods on the optic nerve-fibres are the most important, as being more intimately and radically concerned in the operation of seeing. The motor muscles of the eye are supplied by the third, fourth, and sixth cranial nerves. The third sends fibres to all the muscles of the orbit, except the superior oblique and external rectus. The superior oblique is supplied by the fourth, and the external rectus by the sixth. (For Nerves of the Iris, see Iris.)

The **Arteries** are derived from the long anterior ciliary.

The **Eyelids**, or **Palpebræ**, protect the eye from injury. The upper lid has the levator palpebræ superioris muscle for upward

movements. The tarsus, inside of the lid, is a rigid plate of connective tissue, though fibro-cartilaginous in appearance, and keeps the
lid in form and shape. The *Meibomian* glands, about thirty in
number, on the inner surface of the eyelids, contain cells which secrete a sebaceous matter to prevent adhesion of the lids. The conjunctiva, or mucous membrane on the inner surface, is continuous,
and reflected over a portion of the sclerotica. On the inner edge of
the lids it converges into the lachrymal canal. The *lachrymal gland*
is a flattened body, about the size of an almond, at the superior external angle of the orbit, and secretes the tears by stimulus through
the lachrymal nerve. The orbicularis oculi (circular muscle) contracts the lids. In its paralysis, however, the lid on the affected eye
cannot be closed; and even during sleep it remains open, because the
two muscles that open and close the eyelids are controlled by two
different nerves; *i. e.*, the levator palpebræ superioris, supplied by
the oculomotorius (third) nerve; and the orbicularis oculi, which
receives fibrillæ from the facial (seventh) nerve. In paralysis
of said orbicularis, the lids cannot contract (close); and, in paralysis
of the nerve fibres of said levator palpebræ superioris, the lids cannot be raised (opened), although the movements of the eyeball may
be unaffected and the pupil capable of dilation and contraction as
before. The movement called winking is likewise suspended. Winking of the eyelids may be produced voluntarily; but generally it
takes place involuntarily through reflex action, excited by contact
of air with the cornea, and the accumulation of tears along the edge
of the lower lid. At short intervals this produces an instantaneous
contraction by reflex, through the orbicularis muscle, the edges of
the eyelids being brought together and again immediately separated, in that manner spreading the moisture of the lachrymal
secretion uniformly over the cornea and protecting its surface from
dryness and irritation. In said paralysis the motor power is lost,
but the sensitiveness remains.

The **Accommodation** of the eye is the power of its crystalline
lens to increase or diminish its convexity so as to focus on the
retina the rays of light reflected from objects situated at varying
distances from the eye. An object is seen according to the light
it reflects on the eye. Light itself is not visible. An image is a
picture of an object. To be perfect, the rays from all points of
the object must not intersect, but each ray must arrive at its own
point on the image. That the appreciation of light is by the

13

reflection of rays from an object, is proved by darkening a room and admitting a sunbeam through a minute aperture. The light-rays can be traced by the reflection of the particles of dust floating in the room. If the air in the room be purified from dust, the rays of light cannot be traced; in fact, the part which the sunbeam traverses appears as dark as the rest of the room. The object alone on which the rays strike and reflect is visible. The apparent size of an object is diminished according to its distance from the eye. The power of estimating distance is acquired by habit and improved by practice. With one eye a person may see as far, but not as clearly, as with two eyes, for, with two eyes we see the object from two different points of view, which gives at once a better idea of the proportions of its different parts. That two eyes are required to properly estimate distance is shown when a one-eyed person tries to quickly dip a pen into an inkstand. If we shut one eye and try to place a stopper quickly into the opening of a bottle, the first attempt is seldom a success, for we either overreach or fall short of the mark. The power of estimating distance is greater in persons whose eyes are set far apart and movable from side to side. The accommodation for distance may be changed about three times in a second; but from a near to a distant object it is effected much more rapidly than *vice versa*. The born blind, seeing suddenly, cannot at first judge distance. An infant, not having arrived at the age of reason, or an adult deprived thereof, will reach out his hands to seize the object desired, though it is far away.

At any point farther distant than 65 metres, or about 225 feet (the *punctum remotum*), no accommodation is necessary. Any point from that up to the *punctum proximum*, which is about four inches from the eye, requires accommodation according to the distance. The latter is the limit of visibility for near objects. The range of accommodation is, therefore, from the punctum proximum to the punctum remotum. Accommodation is accomplished through the elasticity of the crystalline lens, which becomes thick or flattened, according to the distance of the object: thick when near, requiring positive accommodation—the anterior surface of the lens becoming more curved (convex); flattened (less curved) when very far off, no muscular action on the lens being then required—*i. e.*, when the ciliary muscle is tranquil, as at rest, the iris relaxes, and the ciliary and suspensory ligaments partially suspend their action, the lens then lying flattened against the vitreous humor behind it. Consequently, in the latter case, near objects cannot be seen properly, while those at a distance can.

A ray of light must traverse the cornea, aqueous humor, lens, and the vitreous humor, before it focuses at the retina. Changes in the figure of the lens are brought about by the iris and ciliary muscles. The circular and radiating fibres of these two muscles contract, thereby exerting sufficient pressure upon the border of the lens to cause the protrusion of its anterior part, which becomes more convex, because at the pupil alone its advance is not resisted. The aqueous humor displaced by the lens enters the anterior chamber, and recedes on the flattening of the lens. The power of increasing the convexity of the lens is diminished by advancement in age. To remedy this defect the person uses convex eye-glasses, to supply the want of convexity of the crystalline lens in accommodation for near objects. Artificial lenses for eye-glasses are generally made of crown or flint glass, in six forms:—

Fig. 56.

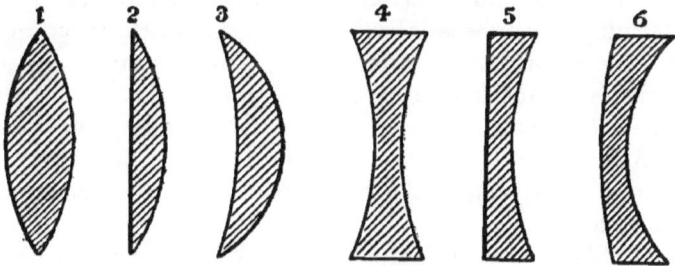

Diagram of artificial lenses.

Thin at edge and thick in middle
{ 1. Double-convex.
2. Plano-convex.
3. Concavo-convex.

Thick at edge and thin in middle
{ 4. Double-concave.
5. Plano-concave.
6. Convexo-concave.

Lenses 1, 2, and 3, which are thickest at the middle, render parallel rays convergent, whichever way they are turned; 4, 5, and 6, which are thinnest at the middle, whichever way these are turned, render parallel rays divergent. The focus, or focal point, is the point at which parallel rays that enter different parts of the lens intersect one another. The more powerful the lens the shorter the focal length. In case of absence of the eye-lens, accommodation is impossible.

Sight.—At the entrance of the optic nerve in the eyeball there are no rods nor cones, consequently no visual sensitiveness. That part of the retina is termed the *blind spot*. The most acute *area of vision* is termed *fovea centralis*, which is a slight depression in the center of the *macula lutea*, directly in the axis of vision, being composed almost entirely of cones, about 2000 in number. This shows that the nerve-fibres are not the part sensitive to light, but that these parts are the cones, forming the terminal organs of said fibres. The rods and cones in the retina lie close together, except in the fovea centralis, which has only cones, showing that they are of more importance to sight. They lie a little deeper than the rods (3 and 4, Fig. 55). When some of the cones are destroyed it causes dark spots of vision. The rods extend nearer to the surface, and are also terminal organs of the optic nerve-fibres. By the rods the quantity of light is determined. The transverse diameter of each rod is about one micromillimetre, that of a cone about two. Impressions on the retina remain from the 1-60th to the 1-30th of a second, experienced by the rapid rotary swinging of a burning torch, especially in the dark, which then appears like a ring, owing to the after-image. On account of the after-image the rapid and involuntary act of winking does not interfere with the continuous vision of objects, as in reading, for instance. The retina, in its adaptation for the reception and selection of the rays of light, makes the eye one of the five organs of the external senses.

Light, like sound, is a quality of the ether accompanied by vibrations, and caused by a body termed in relation to it the "efficient cause." The organ which appreciates this quality (light) is the eye, the organ of sight. The essential nature of light is not known. It is known as a physical fact, however, that light is propagated at the rate of about 186,000 miles in a second, and is capable of transmitting and imparting energy. **Ether** (properly *æther*) is a medium filling all space, within and outside of matter. Its rarity and elasticity are enormously greater than those of air. (*Vide* Æther in chapter on Miscellaneous.) The velocity of light through *ether* is 186,000 miles in a second, while sound travels through *air* only 1-5th of a mile in that time. The transmission of energy through the medium of ether is termed *radiation*. One kind of energy we call light. Its luminous rays on the eye stimulate the sense of sight. Rays of light-ether travel in vibrations, the impulse following by repetition, or periods. The color of bodies results from one portion of the colored rays contained in white light being absorbed at the surface of the body. If that portion which is unabsorbed traverses the body it is colored and transparent; if refracted, it is colored and opaque. Distilled water in a glass appears white; by adding matter which changes the vibrations the white disappears, the color of the matter added being perceptible only. In this case, the increased additional color of one stratum has chemically changed and blended with the other, as well as altered the vibrations. The sensational effect of a stimulus on the external sense-organ does not proportionally affect the internal sense-organ, because the external sense-organ changes the logarithm of the stimulus. The vibrations of light-ether, their repetition and intensity, depend on force. The difference between the colors of the spectrum depends on the wave-lengths of light; or is a difference in frequency of vibration, just as in sound the number of vibrations in a given time determines the pitch of a note. (See Retina.)

Color Blindness may be caused by a failure of certain cones in the retina to absorb light-ether of certain color vibrations, or it may be due to a defect in the area of sight in the cerebrum. This condition is illustrated by a person whose judgment on a certain subject may be entirely abnormal, while on others it is perfectly sound. *Daltonism* depends especially on the inability to distinguish red color.

The **Emmetropic Eye** (normal) conducts parallel rays from distant objects to a focus on the retina without any effort of accommodation of the crystalline lens, or aid on the part of the iris, ciliary muscles, ciliary and suspensory ligaments.

The **Hypermetropic, or Hyperopic, Eye** (short or flat eyeball) (A, Fig. 57) is long-sighted, the rays being abnormally refracted, and focused behind the retina (2). In this case the eyeball is too short antero-posteriorly, though otherwise it may be normal in shape and healthy in function. With parallel rays from far objects the accommodation muscles do not require to be active. A proper convex eyeglass (converging the rays of light) will focus the rays at a shorter distance—that is to say, on the retina (4).

The **Myopic Eye**, short or near sight (B, Fig. 57), must not be confounded with a dim or weak sight, because short-sightedness applies exclusively to the *range*, not to the *power*. The short-sighted eye is optically too powerful, and an object, at a moderate distance, is not properly focused. A person thus affected may possess the acutest power of vision for near objects. If the eyeball is too long, then the rays decussate, or focus within the vitreous body (6), before arriving at the retina, and consequently are not focused on the proper organ (8). The cause of myopia differs in different persons; in some it is caused by an over-convexity of the cornea, or the lens; in others from undue density or overabundance of aqueous humor

Fig. 57.

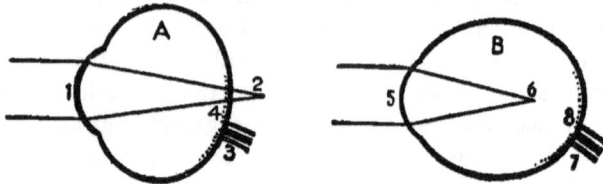

Diagram of a long, and a short-sighted eyeball.

A, Eyeball too short.   B, Eyeball too long.

1 and 5, Anterior portion of eyeball (cornea). 2, Focus behind the eyeball of a long-sighted eye (hypermetropic), the focus having passed the retina (4). 3 and 7, Optic nerve. 4 and 8, Dotted line indicates the retina. 6, Focus within the eyeball, and not reaching to the retina (8) of a short-sighted eye (myopic); the eyeball is elongated.

in the anterior and posterior chambers; from an elongated vitreous body; from a defective accommodation; or, as stated, from an elongated eyeball. The cause is often hereditary in families. Donders believes the prolongation may depend on either one of three causes: "1. Pressure of the eye-muscles, during continued strong attempts at accommodation. 2. Increased pressure of blood in the eyes from a continued stooping position. 3. A congestive process, leading to softening and extension of the eye-membranes." Short-sighted persons require concave glasses to remove the focal point farther back; *i. e.*, on the retina (8, Fig. 57).

Diplopia, or double vision, is of two kinds; caused, first, by a want of harmony in the movements of the eyes, the vision of each, separately, being perfect; second, there may be double vision with one eye only. The first may occur in squinting; the second, from paralysis of one or more muscles of the orbit. The latter form may arise from disease of the brain, cold, or overstudy.

The **Presbyopic,** or long-sighted, eye (A, Fig. 57). The range of accommodation is here diminished, and the vision of near objects interfered with; it is the failure of the power of accommodation in advancing years. This condition is natural from youth to age, like gray hairs, and may occasionally be met with in young persons. This is not so much due to shortness of the eyeball, as is the case with hypermetropic eye. Presbyopia is a defect of accommodation, the lens by age becoming harder, its fibres less elastic, and the ciliary muscle weaker; consequently, the lens becomes flatter, and cannot accommodate itself for distinct vision of near or distant objects. In such a case a remedy may be found in the use of *convex* glasses of low power when reading, but not constantly, their occasional use being not so dangerous as the inconsiderate use of the *concave*.

Hemeralopia, or night-blindness, is a peculiar form of intermittent blindness. The person can see perfectly with an ordinary light, but becomes partially or entirely blind at twilight. This affection is met with in tropical regions, the strong light of which may exhaust the retina, so that it is no longer capable of responding to the weak rays of twilight or moonlight. A remedy is found in protecting the eyes from strong daylight by colored eyeglasses, in tonics, and nourishing diet.

Snow-blindness may be regarded as analogous to the preceding.

CHAPTER XXI.

THE EAR AND THE SENSE OF HEARING.

The organ of hearing is divided into three principal parts — the external, middle, and internal ear. The external ear receives the sound; the middle mechanically transmits it; the internal receives impressions and analyzes their quality, pitch, and intensity. Thus the three, by working together successively and harmoniously, complete, in the peripheral end organ, all that is required, so far as the material part is concerned, for the production of the external sensation of sound.

The **External Ear** (A, Fig. 58) consists of two parts—the *pinna*, or *auricle* (1), and the *auditory canal*, or *external meatus* (2)—which collect, moderate, concentrate, and reflect inward the sounds falling upon them. The pinna, or auricle, is the external skin and cartilaginous framework. Its deep central concavity is the *concha*, and the lowest pendulous part the *lobe*. The irregular surface of the external ear is of some importance to hearing, as, by making the irregularities smooth, with wax or otherwise, though the canal be left free, the intensity of sound and its direction are apparently changed. In such a case the external smoothness reflects the sound-vibrations outwards, and, therefore, diminishes their intensity. The external ear is a sound condenser. Holding the hand in a concave position behind the ear increases the condensation, and, consequently, the intensity of sound. Its three muscles, named *attollens, attrahens,* and *retrahens,* keep it in position, and in man are seldom of sufficient size and strength to move the ear. In some animals these muscles are stronger, not only to gather more sound when listening, but also to relax or open the external part of the meatus to facilitate the entrance of the sound waves. Some animals, such as horses and cattle, have an advantage over man in this respect, as, when the sound-waves are too sharp or painful, or when swimming, they incline their ears backward, thereby closing the meatus.

The **External Auditory Canal,** or *meatus auditorius externus* (2), extends from the concha inward for about an inch, and terminates at the *membrana tympani*. In front of the concha is a prominence directed inward and backward, termed *tragus* (goat), covered with hair, hence its name. At the outer part of the meatus are fine hairs and sebaceous glands, while deeper in the canal are small glands secreting a bitter substance (*cerumen*), called ear-wax, that catches dust and entangles insects, thus preventing their further ingress. When the external meatus is plugged with the secretion (*cerumen*), hearing is impaired; or, if the cerumen is lodged against the membrana tympani, *vertigo*, or dizziness, may be caused. The air in the external meatus, like air enclosed in any tube with an opening, increases sound by resonance; and as all tubes and vacant

Fig. 58.

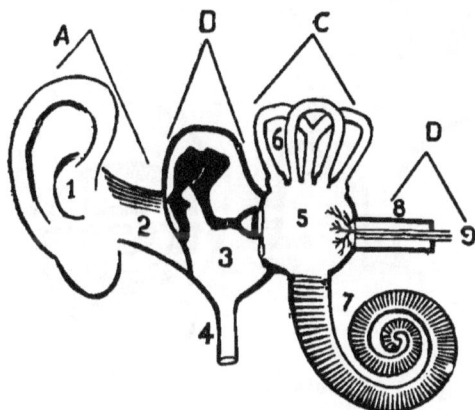

Diagram of the ear.

A, External ear. B, Middle ear. C, Internal ear. D, Internal auditory canal (8), and auditory nerve (9).

1, Pinna, or auricle of external ear. 2, External auditory canal. 3, Middle ear, or tympanum. 4, Eustachian tube. 5, Vestibule of the internal ear. 6, Semicircular canals of the internal ear. 7, Cochlea of the internal ear. 8, Internal auditory canal. 9, Auditory nerve.

spaces, according to size, have different tones when blown into, so the size and length of the external auditory canal modifies the tone. Consequently, it plays a part in the acuteness of hearing in different persons, one considering a certain sound disagreeably shrill, while another considers it very agreeable.

The **Membrana Tympani** (*a*, Fig. 59) slopes obliquely from above downward and inward at the end of the external meatus (2). The slope presents a larger surface to the sound-waves. The vibrations of the membrane depend on its thickness and elasticity. It has no self-tone, as the malleus bone (*b*) in the middle ear (3) is attached to said membrane, acting on it as a damper, and is thrown into the same number of vibrations as the sonorous body. Normally, it vibrates more easily when lax, but when very tense acute hearing is impaired. The handle of the malleus bone, being thus attached, participates in the vibrations, and communicates the im-

Fig. 59.

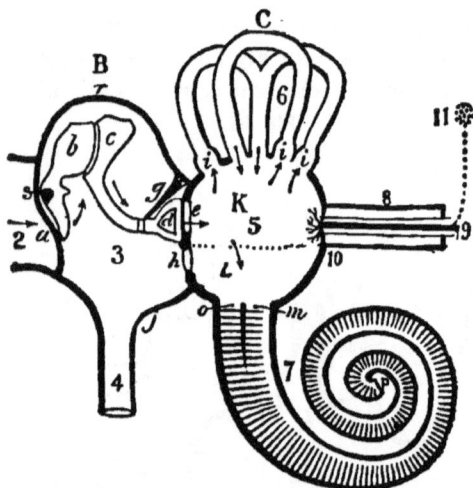

Diagram of the middle and internal ear.

B, Middle ear, or Tympanum.   C, Internal ear.

2, Portion of the external auditory canal. 3, Middle ear, or tympanum. 4, Eustachian tube. 5, Vestibule of the internal ear. 6, Semicircular canals of the internal ear. 7, Cochlea of the internal ear. 8, Internal auditory canal (meatus). 9, Auditory nerve. 10, Fovea hemispherica. 11, Auditory centre in the cerebrum.

a, Membrana tympani. b, Malleus bone. c, Incus bone. d, Stapes bone. e, Fenestra ovalis. g, Stapedius muscle. h, Fenestra rotunda. i, Ampulla of the semicircular canal. j, Floor of the tympanum (middle ear). K, Utricle of the vestibule. L, Saccule of the vestibule. m, Scala vestibuli. o, Scala tympani. r, Roof of the tympanum. s, Tensor tympani muscle.

The dotted line in the vestibule (5) indicates the separating membrane between utricle (K) and saccule (L).

pulse to the incus (*c*), thence to the stapes (*d*) in contact with the membrane of the fenestra ovalis (*e*). The contraction and tension of the membrana tympani, in regulating or responding to sound, may he compared with the regulating power of the iris of the eye. The intensity of sound produces a reflex action, through the nervous system, to the *tensor tympani muscle* (*s*), which then produces more or less tension of the membrana tympani (*a*). The reflex by which the impressions are regulated is effected by a nerve from the trigeminus, through the otic ganglion. When the tensor tympani muscle contracts, the handle of the malleus is drawn inward, producing more or less tension in the membrana tympani; but when it relaxes, the membrane, by its elasticity, returns to its position of inactivity. This variation of tension is its *accommodating power* to receive and transmit modified sound impulses, such as those of pitch, loudness, and quality. The membrana tympani is also the gate guard, permitting nothing to pass into the middle ear that would disturb the delicate mechanical action of the three little bones. The construction of both the external and middle ear is evidently designed for the propagation of sound, and its multiplication by resonance.

The **Middle Ear**, or **Tympanum** (B, 3, Fig. 59), is a cavity, or space, between the membrana tympani (*a*) and internal ear (5), filled with air through the nasal cavity and Eustachian tube (4). It contains three very small bones—the *malleus* (*b*), *incus* (*c*), and *stapes* (*d*)—which, by their articulation with each other, establish communication and transmit sound-vibrations from the membrana tympani to the internal ear. The tympanum, therefore, from its position and office, is called the middle ear. The stapes covers the fenestra ovalis (*e*). The extent of the moderation of oscillation between the membrana tympani and fenestra ovalis may here be noticed. The passage of the oscillations through the three little bones, and the difference in size between the membrana tympani and the fenestra being as thirty to one, the extent of movement of the former is much modified; otherwise, it would burst the latter. The modification is effected by the leverage of said bones, the vibrations losing in extent but gaining in force. The manner in which sound is propagated through these bones is due partly to molecular action and partly to the movements of the same bones as a whole. The middle ear is lined with epithelium, being a continuation of that of the nose, through the Eustachian tube. In the wall between the middle and the internal ear are two openings, or windows, termed fenestræ (*e* and *h*), each lined with a thin membrane. The larger is termed *fenestra ovalis* (*e*), leading to the vestibule of the internal ear, and is covered by the base of the stapes (*d*). The smaller opening is termed *fenestra rotunda* (*h*), and communicates with the cochlea. Both fenestræ lie in contact with the fluid endolymph in the internal ear (5).

The middle and internal ear are situated within the temporal bone, which at that part is thick and hard—the *petrous* portion—so that the delicate and complicated structure inside may be the more fully protected. The space (3) of the middle ear is filled with air, thereby giving sound an acoustic effect. Its roof (*r*) separates it from the brain. The bone here is a thin plate, through which, if defective, inflammation of the brain may occur in case of abscess in the middle ear. Its floor (*j*) is also a thin plate of bone, separating it from the jugular vein and carotid artery. Any accident to the head may burst the thin plate of bone and rupture the artery or vein, producing hæmorrhage into the middle ear, the blood escaping through the mouth, nose, and possibly through the external ear. Again, a person with a perforated membrana tympani may drown in water, though he be a good swimmer, the water entering the middle ear, and through the Eustachian tube into the throat.

The **Malleus Bone**, or hammer (*b*, Fig. 59), consists of a head, neck, and three processes. The head articulates with the incus (*c*), the neck continuing to these processes. The first of these, termed *gracilis*, is thin, small, and long, to which is attached the levator tympani muscle, affording leverage to facilitate the movements of this bone; the second, termed *brevis*, is pointed, as is the rest of the manubrium, and affords attachment for the tensor tympani muscle, lying in contact with the membrana tympani, pressing it slightly outward (concave), and acting as a damper in preventing after vibrations. When waves of sound are too powerful and the damper not sufficient, then strong vibrations may produce pain sympathetically. The third process is long and pointed, termed *manubrium* or handle, attached to the membrana tympani, and taking part in its vibrations. The head of the malleus articulates with the incus by a loose joint, so that when its handle moves correspondingly with the vibrations of the membrana tympani the long process of the incus does not move therewith to that extent, as it is only two-thirds as long as the handle. Besides, the latter moves farther from its axis, and, therefore, the extent of vibratory movements on the stapes is considerably modified. They are, however, more powerful in its base on the membrane fenestra ovalis, in its movements to and fro.

The **Incus Bone**, or anvil (*c*, at B, Fig. 59), consists of a body and two processes. Its body articulates loosely with the malleus bone, and its long process with the head of the stapes, and its short one is attached to the margin of the opening leading to the mastoid cells, or cavities formed in the mastoid process by absorption of bony tissue.

The **Stapes,** or stirrup (*d*, Fig. 59), consists of a head, neck, two branches, and a base. The head articulates movably with the long process of the incus; the neck receives the stapedius muscle (*g*), the smallest in the body, originating just behind the fenestra ovalis, and holding the stapes in position on the membrane covering said fenestra. The two branches, or *crura*, of the stapes terminate at its base. The presence of three bones instead of one in the middle ear is to moderate the vibrations of the membrana tympani, so that said impulse of vibration may not injure the more delicate nervous structure in the internal ear and brain. These bones are covered with a membrane derived from the wall of the middle ear, and serve a twofold purpose—as a transmitter of sound, and for attachment of muscles. Adhesion, or anchylosis, at one or more articulations of the malleus, incus, or stapes, causes difficult hearing.

The **Muscles of the Middle Ear.** — The *tensor tympani muscle* (*s*, Fig. 59) produces tension of the membrana tympani inward, its contraction not being continuous, but varying as the sounds differ, and by itself not under the control of the will. Normally, it is excited reflexly through nerve impulses, similar to the iris of the eye. The *laxator tympani muscle* draws the malleus bone outward, relaxing the membrane. Both muscles are attached to the malleus, the tensor (*s*) to the manubrium, the laxator to the process gracilis. The *stapedius muscle* (*g*) originates at the eminentia pyramidalis, just behind the fenestra ovalis (*e*), and is inserted in the neck of the stapes, the base of which is thereby depressed and held in position on the fenestra ovalis, which is surrounded by an annular ligament, keeping the stapes in a more fixed position. The stapedius muscle, when active, moderates or prevents severe vibratory movements of the incus on the stapes. The nerve filaments of these muscles are from the facial nerve.

The **Eustachian Tube** (4, Fig. 59) is about one and a half inches long. It establishes communication between the middle ear and nose. Disease in the nasal cavity may partially or entirely destroy the function of this tube for a time; for instance, by chronic catarrh, by thickening of the mucous membrane, pressure by cicatrix or by tumor, producing partial or total deafness, either temporarily or permanently, the air in the middle ear vibrating then with difficulty, or perhaps not at all. In some cases the obstructive membrane may be remedied by the introduction of a Eustachian catheter (O————⌐) through the nasal cavity, or of air with the Politzer's bag. When the bag is used its nozzle is placed in the nostril, and while pressing the bulb and forcing the air into the middle ear, the effort to swallow must be made. The nasal orifice of the Eustachian tube is normally closed; but in the act of swallowing, or in other motions of the pharynx, it opens, and hence arises the necessity of swallowing during the injection of the air. When too much air remains in the middle ear hearing is dull, as it unduly distends the membrana tympani, even to the extent of causing pain. The act of swallowing, repeated a few times, will relieve this trouble if the tube be not compressed as above mentioned. The Eustachian tube is intended as an air-passage and outlet for mucus from the middle ear. In the effort at expiration, with the mouth and nose closed, air may be driven into the tympanum (middle ear), while the effort at inspiration under the same conditions withdraws it from that cavity, the air passing in and out through the Eustachian tube connecting the pharynx with the tympanum, regulating thereby the pressure on the membrana tympani. Permanent occlusion of the tube is a common cause of deafness.

The **Internal Ear**, or labyrinth (5, 6, 7, Fig. 59), consists of passages hollowed out within the thickest part of the petrous portion of the temporal bone, and is termed the *osseous labyrinth*. The membranous lining inside is called the *membranous labyrinth*. This latter, by its cavities and tubes, forms the vestibule, cochlea, and semicircular canals—all joining, and constituting in their entirety a closed sac in the internal ear. This membranous labyrinth is separated from contact with the surrounding bony structure by the *liquor* Cotunnii, a limpid fluid termed *perilymph* (14, Fig. 60). This fluid is secreted by cells in the wall of the membranous labyrinth, and is finally carried away through the *aqueduct cochlearis* and taken up by the lymphatics.

The internal ear is the most essential part of the organ of hearing. Its cavities and passages are divided into three portions, the vestibule (5, Fig. 59), semicircular canals (6), and cochlea (7), communicating with each other directly, and outwardly with the middle ear through the fenestra ovalis (*e*) and fenestra rotunda (*h*), also with the brain by the auditory nerve (9), which passes through the internal auditory meatus (8).

The **Vestibule** (5, Fig. 59) is an irregular ovoid chamber, or central cavity, filled with endolymph, and is about one-fifth of an inch in diameter. It communicates not only with all parts of the internal ear, but outwardly, by the *fenestræ ovalis* and *rotunda (c and h)*, with the middle ear (3), and inwardly, by the *fovea hemispherica* (10), with the internal auditory meatus (8). The wall or membranous lining of the fovea hemispherica has many perforations, termed *maculæ cribrosæ*, for the passage of nerve filaments of the auditory nerve (9). The fenestræ of the middle ear form the only movable separating medium between the air contained therein (3) and the fluid in the labyrinth (5, 6, 7). The membrane covering the fenestra ovalis receives vibratory impulses from the stapes, which are communicated to the endolymph in the internal ear. The sonorous stimulus for oscillations of the endolymph may also be produced through the bones of the skull, by either placing a sounding body on the head, or a tuning fork between the teeth, when the ears are closed. In deafness, a sounding fork between the teeth may assist in diagnosis; if sound be perceived, it indicates that the internal ear, or labyrinth, is in good condition, in which case the trouble will likely be found in the middle or external ear. The vestibule communicates with the semicircular canals (6) and cochlea (7), and is divided into two sacs (indicated by dots) the utricle (k) and saccule (L); the utricle being the largest, and receiving the membranous semicircular canals (6) by five orifices. It also communicates by a small Y-shaped tube with the *canalis reuniens* forming the ductus endolymphaticus that passes through the osseous aqueduct, which terminates in a blind sac in the subarachnoid space. The saccule (L) is the smaller sac communicating by a small tube with the canalis reuniens and ductus endolymphaticus. The utricle (k) and saccule (L, Fig. 60) have each an oval area termed *macula acustica* (12), containing small bodies termed *otoliths*—crystals of carbonate of lime (12)—lying in the viscid endolymph in the utricle (k) and saccule (L). The otoliths are intermixed with *otoconia*, or ear powder—crystals more minute than the otoliths. The use of the otoliths and otoconia is to strengthen and divide equally the sonorous impulses in the endolymph; that is to say, they play the part of dirigents. If the sound makes a fine, gentle, and pleasing impression on the contents of the vestibule, then the impulse will be sent to the cochlea (like a *gentleman* into the parlor), where musical tones and enjoyable harmony results. Should the sound make a harsh, disagreeable impression, then it is relegated to the semicircular canals, resulting in disturbance and confusion. If the sound be of a mixed character, as from a wagon passing over a cobble-stone street, it is a *noise*. The vestibule, therefore, may be described as an enclosed porch, or hall (Vorhof), where all sound impulses are directed to their respective compartments, according to quality. The nerve-fibrillæ of the auditory nerve (9) supply the utricle (k), saccule (L), and ampullæ (i), and also help to hold the membranous labyrinth in position. The wall of the membranous labyrinth consists of three layers—the outer, middle, and inner. The latter is formed of epithelial cells secreting the *endolymph* (liquor Scarpæ), a limpid serous fluid which fills all the cavities and tubes of the membranous labyrinth, such as

Fig. 60.

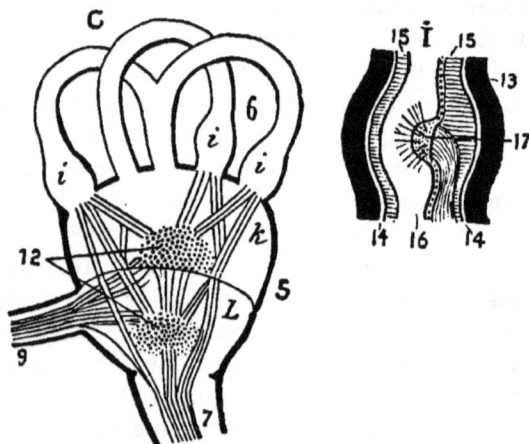

Diagram showing the distribution of the auditory nerve-fibrillæ within the vestibule.

C, Portion of the internal ear. *i*, Ampullæ of the semicircular canals. *k*, Utricle of the vestibule. L, Saccule of the vestibule.

5, Vestibule. 6, Semicircular canals. 7, Leading to the cochlea. 9, Auditory nerve. 12, Otoliths, and Otoconia.

I, One ampulla enlarged. 13, Bony wall. 14, Space between the bony and membranous walls, occupied by perilymph. 15, Wall of the membranous canal. 16, Internal portion of the membranous canal, occupied by endolymph. 17, Crista acustica having auditory nerve-fibrillæ projecting into the endolymph within the canal.

the vestibule, semicircular canals, and cochlea. The endolymph is the substance in which the true nerve-filaments pass to and fro (Fig. 60), and into which the nerve endings for the sensation of hearing project (Fig. 62). It is renewed by the epithelial secreting cells; the worn-out lymph and cells passing off through the arachnoid layer of the auditory nerve (9, Fig. 60) into the subarachnoid space of the brain.

The **Semicircular Canals** (6, Fig. 60) are three *osseous* hollow windings surrounding the *membranous semicircular canals.* These canals are each about 1-20th of an inch in diameter, and terminate in the utricle (k) by five orifices, one of which is common to two of the canals. Three of the apertures have enlargements at one end, termed *ampullæ* (*i, i,*). Each ampulla has a projection, or tubercle, termed *crista acustica* (17), into which auditory nerve-fibrillæ enter, and project, as it were, into the endolymph within the canal (16). These terminations of the nerve-fibrillæ may properly be termed *equilibrium organ*, as they regulate the equilibrium of the head, and its rotation as to rate and direction. A lesion of the horizontal semicircular canal disturbs this equilibrium, producing alternate lateral movements (right and left) of the head, often noticed in persons of advanced years, though hearing may not be interfered with. A lesion of the vertical semicircular canals causes up and down (nodding) movements of the head. Disturbance of all three semicircular canals produces loss of equilibrium in many directions (such as that in which the eyeball and lids participate), when the person may not be able to stand still, and may even fall. Such a disturbance may also be caused by weakness in the reflex centre of origin of certain auditory fibres in the cerebellum. In one instance lukewarm water gently injected into the ear of a healthy man of sixty to remove earwax produced such an effect on the nerve-fibrillæ projections of the crista acustica that it caused him to stagger and fall on three different occasions, and blinded him for three or four minutes. The attempt to remove the wax with the curved end of a lady's common hair-pin resulted similarly.

When the eyes are bandaged and the body quickly rotated, equilibrium may be preserved for the first few turns, but it is soon lost, and a false impression of the relation of the person to his surroundings takes place. In such a case the faculties of ascertaining proper position are either partially disturbed or entirely lost, and *vertigo*, giddiness, or dizziness, ensues. The revolving movement affects the endolymph in the ampullæ of the semicircular canals, causing a disturbance on the nervous hair-like end projections in the ampullæ, by moving or floating them out of normal position. Giddiness

14

may also be produced by impressions on the eye, for the cerebral area of sight and that of hearing are both in communication with the *otic ganglion*, and this, with the eye, corpora quadrigemina, and the membranous ear labyrinth. This accounts, too, for the loss of equilibrium and falling down, sometimes, of a person on suddenly beholding a shocking spectacle, as well as for the sympathetic connection—through the otic ganglion—between the membranous labyrinth and the eye, in certain diseases of either ear or eye. (*Vide* Auditory Nerve.)

The **Osseous Cochlea** (7, Fig. 59; and Fig. 61) resembles a snail-shell. It is about a quarter of an inch in diameter at its base

Fig. 61.

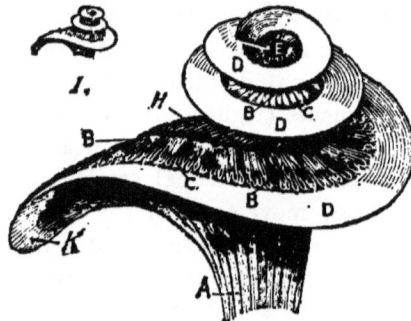

Diagram of the cochlea.

A, Nerve-fibrillæ of the auditory nerve passing from the sacculus portion of the vestibule to and within the modiolus of the cochlea. B, Nerve-fibrillæ emerging from the modiolus through the osseous spiral lamina (H). C, Terminations of nerve-fibrillæ at the organ of Corti. D, Shape of the membranous cochlea. E, Helicotrema at the cupola of the cochlea. H, Indicating the osseous spiral lamina. K, Commencement of the scalæ. 1, Natural size of the cochlea, about 1-4th of an inch in width and height.

(1, Fig. 61), and forms a spiral tube winding two and a half times around a central pillar termed the *modiolus*, and constitutes the anterior portion of the labyrinth, or internal ear. The central bony portion of the modiolus has minute channels for the passage of nerve-fibres and arteries (*arteria centralis modioli*). Both the modiolus, which terminates at the cupola (E), and the canal of the cochlea diminish in diameter from the base toward the apex, the canal being about half an inch in length. All along the modiolus is a projecting thin plate of bone termed *osseous spiral lamina* (6, Fig. 62), projecting about halfway into the canal, following its windings and partially dividing it into two parts, or chambers (indicated at *o* and *m* Fig. 59, but more plainly at the transverse section of the one-half winding of the cochlea B, Fig. 62). The osseous spiral lamina and the membrane of Reissner (1), and basilar membrane (3), subdivide the membranous canal of the cochlea into three canals, termed *scalæ*; one of which (*o*, Fig. 59) is termed *scala tympani* (A, Fig. 62), communicating with the fenestra rotunda (*h*, Fig. 59). In fact, it would open into the middle ear if the fenestra rotunda were not closed by a membrane. Another (*m*, Fig. 59) is termed *scala vestibuli* (A, Fig. 62), communicating with the vestibule (K, 5, Fig. 59). These two scalæ also intercommunicate at the apex, or cupola of the cochlea (E, Fig. 61) through a small opening, the *helicotrema*. A third canal, or scala, is formed as follows: From the osseous spiral lamina (B, Fig. 62) three mem-

Fig. 62.

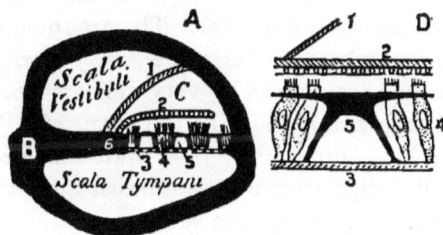

Diagram of the scalæ and organ of Corti in the cochlea.

A, Transverse section of a one-half turn of the cochlea. B, Modiolus, or central stem of the cochlea. C, Scala media, or ductus cochlearis. D, An enlarged section of the organ of Corti.

1, Reissner's membrane. 2, Membrane of Corti. 3, Basilar membrane. 4, Cells of the organ of Corti. 5, Osseous laminæ spans supporting the cells of the organ of Corti. 6, Osseous spiral lamina and indication of auditory nerve-fibrillæ passing from the modiolus (B) through the osseous spiral lamina (6), to the cells of the organ of Corti (4).

branes arise (1, 2, and 3), which gradually separate and cross the canal. The upper is called *Reissner's membrane* (1), forming the roof of the third canal (C), termed *scala media*, or *membranous ductus cochlearis*; the middle membrane (2), which stretches through the greater portion of the scala media, and is called *membrane of Corti;* the lower membrane (3), forming the floor, is termed *basilar membrane* (*Vide* Membrana Basilaris). The scala media, or ductus cochlearis (C), is about 1-10th of an inch in diameter, and is of a triangular form. It is separated from the scala vestibuli by Reissner's membrane, and from the scala tympani by the basilar. The membrane of Reissner not only modifies the movement of the endolymph in the scala vestibuli, but also purifies the endolymph; for the ductus cochlearis (C) is filled with endolymph, which passes from the scala vestibuli through Reissner's membrane by endosmosis, so that the ciliæ (rods of Corti) float in the purest endolymph. The three scalæ so formed—*i. e.*, the scala vestibuli, scala tympani, and the ductus cochlearis—are each filled with endolymph.

The **Membrana Basilaris** (3, Fig. 62) consists of fibrous bands stretching across from the osseous spiral lamina (B and 6) to a point termed *crista basilaris* on the bony wall opposite. The inner surface of this membrane is beset with about 3000 cells (4), each having ciliated (hair-like) projections, termed *rods of Corti.* The cells and their rods are remarkable in arrangement, running in a series from large to small, like pipes of an organ, as the ductus cochlearis diminishes in diameter. The grouping of cells which rest on the basilar membrane is termed *organ of Corti.* These cells are supported on this membrane by very delicate bony tissue laminæ stretching across with it (5), like the spans of a bridge, giving support to the cells, in order that their hair-like projecting ciliæ can be more readily impressed by the number and extent of vibratory impulses received through the endolymph in the scala media (C), into which they project. These ciliated projections (rods of Corti) are the external sense-organs of hearing. The nerve-fibres are sensory and belong to the auditory nerve (*portio mollis*). At the vestibule, the saccule fibres of this nerve (L, Fig. 60) enter and follow the minute bony canals within the modiolus, and issue therefrom at regular intervals, passing along within the osseous spiral lamina, as indicated at (6, Fig. 62), and then enter the cells of the organ of Corti. The rods of Corti on the cells (4) have the power of responding to sounds of different pitch, according to the number of vibratory impulses passing through the endolymph of the ductus cochlearis (C), and are, therefore, the true external sense-organs of hearing. The long, thick rods at the base of this duct respond to deep bass notes, and the thin short rods at its apex to those of higher pitch.

Sound is a quality perceived by the hearing arising from a certain species of vibrations in the air, or in any other mass capable of being properly excited, and is caused by a sonorous or sounding body. Sound, therefore, is in the medium itself (be this medium a gas, fluid, or solid) that conveys it to the organ of hearing. A

Vibration is a change of position of particles of matter in quick succession; its equal interval of time being called the *period* of vibration. In the same proportion that an elastic medium facilitates the velocity of vibrations a dense one retards it, so that the greater the elasticity the greater the facility and rapidity of transmission of condensed and rarefied waves, or vibrations, through such medium. Again, the intensity of sound varies inversely as the square of the distance; so that a sound which has a certain intensity at a unit of distance will have only one-fourth the intensity at double the distance. Intensity likewise depends upon the amplitude of the vibrations, the density of the medium in which the sound is produced, the direction of air currents, and the proximity of other sonorous bodies. A **Wave-length** is the distance between the two nearest particles of a medium in the same phase of vibration. This length bears a fixed ratio to the vibrations produced by the sounding body. The greater the number of vibrations per second the shorter the wave length. Sound travels through air at the rate of about 1100 feet in a second (about 1-5th of a mile), while light travels about 186,000 miles per second. The extreme limits of regular periodic vibrations of sound (tones) is between 30 and 30,000, but beyond these limits they cannot be perceived. Sounds have certain characteristics by which they are recognized as either musical tones, or noise. When vibrations arrive at the ear with regularity of intensity, periodicity, and simplicity, a musical tone is the result, having three essential characters—loudness, pitch, and quality. Irregular and rapid non-periodic vibrations are the sounds called *noise*. Sound waves are refractive, as shown by the celebrated experiment of Sondhauss; and reflective, as beautifully illustrated in whispering galleries, especially in those of St. Paul's, London, and of the famous gallery in the Mormon Temple at Salt Lake. The latter is elliptical, the slightest whisper at one focus being faithfully reproduced at the other. Musical sounds differ only in loudness, pitch, and quality. **Loudness** (intensity), as before stated, depends on the amplitude of the vibrations; *amplitude* being the distance, or height, of the crest of a wave from its axis (the imaginary central line drawn longitudinally through the wave). **Pitch** is the difference between a low and a high tone, depending on the number and rapidity of vibrations impinging on the ear in a second. Any two instruments giving forth tones of the the same pitch must produce the same number of vibrating wavelengths in a given time; the more rapid these are, the higher the pitch. **Quality**, or, as it is also called, *timbre* or *sound-clang*, is dependent on neither pitch nor intensity, but upon the form or *character* of the vibrations. It is the peculiarity of tone which enables the listener to determine whether a tone comes from a violin, flute, zither, harp, piano, organ, bell, etc. All these different instruments, in order to produce the same tone, must give off the same number of vibrations; at the same time, each has a peculiar quality, sound-clang, or timbre, by which one instrument may be distinguished from another. It depends really on what are called overtones. Overtones are sounds produced by a body vibrating *in parts*, these auxiliary vibrations being superimposed upon and accompanying the fundamental tone produced by the vibration of the body *as a whole*. These compound wave-vibrations, striking the membrana tympani (in a manner somewhat similar to that experienced on the finger from a dicrotic pulse), produce a sensation quite different from that produced by the fundamental tone alone. To pick out or distinguish these overtones requires generally the trained ear of the musician. As, however, the mechanism of the organ of Corti is not the same in all individuals, the nicety of discrimination of these overtones, be they manifested as harmonious or not, must vary in a corresponding degree.

Resonance is the strengthening and prolongation of sound produced by reflection or by reinforcement received from the vibrations of other bodies. If it be caused by reflection, the reflecting surface must be less than 112 feet distant from the sounding body, so that the direct and reflected sounds are blended. They cannot be heard separately, though they are mutually strengthened. If the distance between the sounding body and reflecting surface be such that there is an interval between the reflected sound and the direct sound, we have what is called an *echo*.

Sound waves occur in different ways, such as by the vibration of the *air* around a vibrating bell; a membrane, as the membrana tympani; through a hollow tube; through a *solid*, as the ossicles (bones) of the middle ear; or, through a *fluid*, as through the endolymph in the internal ear. Sound may cause an impression on the ear by a single wave, illustrated by a pistol-shot, or by a continuous succession of waves, as by a bell or piano-string. Sound waves passing through a tube do not disperse; consequently, they are transmitted to a greater distance with little loss of force. The sound motion of air particles is backward and forward in the line of propagation; not up and down, across that line, like water waves.

Hearing.—The fluid in the auricular labyrinth may be set in motion by sound-vibrations in two different ways: 1. By the base of the stapes in contact with the fenestra ovalis; and 2. by vibratory motion communicated through the osseous walls of the labyrinth. Ordinarily, it is effected through sound-waves entering the external meatus and impinging upon the membrana tympani. The vibrations are thence transmitted through the malleus, incus, and stapes of the middle ear to the membrane covering the fenestra ovalis, lying in contact with the endolymph of the utricle in the vestibule of the internal ear. The endolymph thus receives the impulse. Pressure on one side of a bladder filled with liquid necessarily causes its other sides to bulge; but the membranous labyrinth cannot yield, as it is surrounded by the osseous labyrinth. The movement, therefore, continues through the endolymph of the scala vestibuli(m, Fig. 59), and (A, Fig. 62), thus causing Reissner's membrane (1) to vibrate correspondingly; and producing more or less modified vibration of the endolymph in the ductus cochlearis (scala media, C). Again, the scala vestibuli, being in communication with the scala tympani at the cupola of the cochlea through the helicotrema (E, Fig. 61), the vibration of the endolymph extends through the scala tympani, which communicates with the fenestra rotunda, whose elastic membrane yields, and thus finally compensates for the bulging movements of the membrane covering the fenestra ovalis. The vibration of the endolymph of the scala tympani causes the membrana basilaris (3, Fig. 62) to vibrate correspondingly, thereby causing a more or less modified vibration of the endolymph on the opposite side of the ductus cochlearis (C). It may here be noticed that two organs—the membrane of Reissner (1), and the basilar membrane (3)—act as transmitters and modifiers of sound vibrations in the endolymph of the ductus cochlearis, which two functions are facilitated by the elasticity of the membrane covering the fenestra rotunda (h, Fig. 59). Thus, vibrations of the endolymph within the ductus cochlearis (C) are greatly modified, and sound is there resolved into its three properties of intensity, pitch, and quality. The ciliæ, or rods of Corti, projecting into the endolymph of the ductus cochlearis, are thus enabled to become impressed. Each cell and its projecting rods of the organ of Corti differs in length and delicacy. Each rod of the three thousand is the terminal of an auditory nerve-fibrilla, and has the function of receiving the impression of a certain number of vibrations, the impulse of which is transmitted to the auditory cen-

tre in the cerebrum. To stimulate the terminals of the auditory
nerve the sound vibrations must have certain qualities—*i. e.*, sufficient amplitude and duration; and, to excite a sensation of a continuous musical sound, a certain number of vibrations must arrive
at the ear in a given time—say, in one second. The membrane of
Corti (2, Fig. 62) projects into the endolymph of the ductus cochlearis, and, when in repose, almost touches the rods. During its
own vibration, however, it acts as a damper, permitting several or
many rods to receive impressions more or less simultaneously. The
cells of the organ of Corti (4) rest solidly on the basilar membrane
(3), and their several rows of rods, or hair-like ciliæ, are supported
by the osseous spans (5). Now, if we suppose that a fundamental
note or tone, say C, with 256 vibrations in a second, is appreciated
by the appropriate rod, then its quality, or timbre, being a compound of vibrations, must be appreciated by as many different rods
as there are differences of vibration in the compound. In this manner is established with facility impressions of intensity, pitch, and
quality. The rods thus impressed, analyze the sound; a nerve impulse is transmitted to the auditory centre, and the external sensation of musical sound follows. The auditory area, or centre, is at
the first temporal or temporo-sphenoidal gyrus (E, Fig. 49), the
destruction of which produces deafness of the opposite ear. The
destruction of the entire gyrus in both hemispheres produces what
the Germans call *Seelentaubheit*—deafness of spirits (soul-emotions).
The sensations of sound, light, and other external sensations, do
not take place in their respective areas, or centres, in the cerebrum,
but, as we have shown, in the external organ itself. Normally,
the sense of hearing may be improved in acuteness and delicacy
by practice, as is illustrated in the experiment of the piano-tuner
and musician.

CHAPTER XXII.

THE VOICE AND SPEECH.

Voice, or the production of vocal sound, is a voluntary act; that is to say, it is under the control of the will. When tone is produced, air is expelled from the lungs through the air tubes and rima glottis with greater force. The *rima glottis* (1, Fig. 63), which separates the vocal chords (4, 4), is a small oblong aperture, wider posteriorly than anteriorly.

Physically, for articulation, eight small yet distinct muscles are involved in each lateral half of the throat within the thyroid cartilage, five of which act on the vocal chords and three on the epiglottis. The former muscles are the thyro-arytænoideus, crico-thyroid, crico-arytænoideus posticus, crico-arytænoideus lateralis, and arytænoideus; the latter are the thyro-epiglottideus, arytæno-epiglottideus superior, and arytæno-epiglottideus inferior.

The *thyro-arytænoid* muscle, one on each side (2, Fig. 63) is the principal agent in drawing the thyroid cartilage forward and down-ward, and bringing the vocal chords (4, 4) nearer to each other. At the same time, these muscles swell inwardly, rendering the chords (the extreme edge of the mucous membrane of these muscles) tense. When the glottis is nearly closed, the *crico-thyroid* (6) and the *posterior crico-arytænoideus* muscles (8) cause the margin of the chords to become concave. At the same moment the contraction of both thyro-arytænoid muscles (2) effect a straightening of the extreme concavity of the chords (4, 4), which become tense and are set in vibration by the air passing outward. At first, a little air escapes without producing sound; then, at the margin of the concavity, two grayish-looking membranes (4, 4) appear stretched from end to end, and a sound, or tone, is produced. These two membranes (the *chords*) are just opposite each other, and are simply extensions of the mucous membrane on the anterior edge. Stretching makes them more grayish, hard, and solid. During their vibration they do not exactly touch one another. The muscular fibres of the membrane are extra strong. The thyro-arytænoid muscle (2) does not vibrate, but it effects the tension of the membrane on its anterior edge. The posterior crico-arytœnoid muscles (8, 8) dilate the glottis. The *crico-arytœnoid lateralis* muscle (7) is inserted in the anterior margin of the arytœnoid cartilage (9, 9), and acts to draw it forward. Paralysis of the latter muscle renders the voice harsh and deep, on account of the insufficient tension of the chord resulting; but paralysis of the thyro-arytœnoid muscle (2) causes a total loss of the voice. The *arytœnoideus* muscle (11) is inserted into both posterior arytœnoid cartilages (9, 9), holding them together.

Fig. 63,

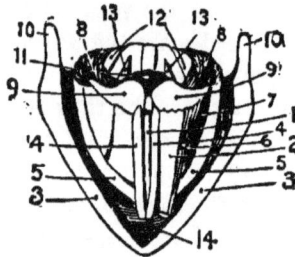

Diagram of the vocal chords and surrounding organs.

1, Rima glottidis. 2, Thyro-arytœnoideus muscle. 3, Thyroid cartilage, constituting the POMUM ADAMI. 4, Vocal chords. 5, Cricoid cartilage. 6, Crico-thyroid muscle. 7, Crico-arytœnoideus lateralis muscle. 8, Crico-arytœnoideus posticus muscle. 9, Arytœnoideus cartilage. 10, Superior cornu of the thyroid cartilage. 11, Arytœnoideus muscle. 12, Posterior continuation of (5), or cricoid cartilage. 13, Capsular ligaments connecting the arytœnoid and cricoid cartilages. 14, Superior thyro-arytœnoid ligament.

The thyro- and the two arytœno-epiglottidei muscles are not shown in the diagram.

The *thyro-epiglottideus* muscle arises anteriorly at the inner surface of the thyroid cartilage, just anterior and external to the thyro-arytœnoid muscle, and extends to the outer portion of the epiglottis. During vocalization it partially regulates the passage of air through the aperture by depressing the epiglottis. The *superior arytœno-epiglottideus* muscle takes its rise from the apex of the arytœnoid cartilage, and extends to the layer of muscular fibres of the mucous membrane on the side of the epiglottis. During vocalization it assists in regulating the passage of air by constricting the superior aperture of the larynx. The *inferior arytœno-epiglottideus* muscle takes its rise from the inner lateral portion of the arytœnoid cartilage just above the attachment of the superior portion of the vocal chord, and extends to the inner and upper part of the epiglottis. During vocalization this muscle assists in regulating the passage of air by compressing the sacculus laryngis.

When the tones of speech are produced, the air in contact with the vocal chords becomes alternately condensed and rarefied by the vibration of the chords, which oscillate rapidly, like the strings of a piano. Hence it is the condensed and rarefied air that sounds, and not the chords or string. (*Vide* Sound in chapter on Hearing). The cavities of the larynx, pharynx, nose, mouth, and trachea form a compound resonating-tube. The muscles connected with these cavities are acted upon through nerve-fibres, and in their action adapt themselves to the sounds and the pitch of the tone produced by the vocal chords. If one or more of these cavities or muscles be in an abnormal condition, resonance is imperfect. We can act at will on the vocal muscles to produce resonance, even below the chords; thereby producing a chest tone. In this the chords are not brought close together, and the vibration is communicated to, and perceptible at, the chest-wall; or, we can produce a head note (falsetto), which is accomplished by causing the vocal chords to come closer together. Hence, the resonance in the trachea ceases, though in the posterior nares it is more pronounced. The limit of the human voice lies between 100 and 1000 vibrations per second, embracing about three octaves; though an expert in musical sounds may appreciate tones ranging from 30 to 38,000 vibrations per second. The note C of the middle octave on the piano vibrates about 256 times in a second.

The pitch of the tone depends on the tension of the marginal membrane of the chords, which may be likened to the stretching of the strings of a violin; but when the chords are brought in contact sound ceases. Again, the length of the vocal chords affects the tension, and so affects the pitch of the tone given forth. Men with a large thyroid cartilage (*pomum Adami*) have deep bass voices. The vocal chords of the female are about one-third shorter than those of the male. At the age of puberty the vocal organs grow rapidly, especially in the male, and the voice changes to a deeper bass tone. External pressure on the glottis causes a different sound to be emitted, even with the same pressure of air. Again, if the epiglottis be entirely open, no sound is produced, even with the strongest force of air; but as soon as the chords come close together and air is forced between them, sound emanates. The closer the chords the less air is required to produce a tone of high pitch;

but the looser and wider apart, the more air is required. Hence, a singer can execute an effective *crescendo* much easier, and without extra inhalation, in the higher notes than in the deep bass ones. In some individuals, when the chords are separated to the extent of about 1 millimetre, deep bass tones can be produced, but when farther apart no tone can issue. Voice is normally formed always in expiration. When a vocal sound is to be produced the chords are suddenly made tense and brought close to each other; the air is driven by forcible expiration between them, and the sound comes forth. The tone, pitch, and intensity of sound vary with the degree of tension of the chords and the force of expiration.

Vocalization is accomplished in the larynx, but articulation is effected by the lips, tongue, teeth, and palate. **Articulation** is the utterance and joining together of the elementary sounds of a language by the appropriate movements of the vocal organs. It is regulated by efferent impulses through the facial and hypoglossal nerves, while vocalization is regulated by efferent and afferent impulses through the vagus—its superior laryngeal branch being the sensory (afferent) nerve of the mucous membrane of the larynx. The vagus also supplies efferent (motor) fibres to the crico-thyroid muscles. All the other muscles employed in phonation are supplied by efferent fibres of the inferior laryngeal nerve.

Speech may be defined as the utterance or articulation of words (sounds) to manifest concepts. Evidently, concepts may be manifested by gestures, nods, etc.; but more effectively by the voice (articulation). The **Speech-centre** is at the inferior portion of the third frontal convolution, near the bottom of the fissure of Sylvius, in the left cerebral hemisphere. If, in the case of speech, the sensory and motor nervous centres, which must necessarily co-exist, as shown above, for co-ordinate action, become vitiated by disease or impaired from any other cause, speech must be defective. If the former centre is affected it produces *sensory aphasia*, there being a deficiency or suspension of the power of recalling the word, or of prompting the necessary nervous actions to pronounce it; if the latter, there is *motor aphasia*, or loss of power to produce the proper muscular co-ordination necessary for articulation of speech. In some cases, a third, and even other motor centres are called into action. For instance, an organist having received a stimulus, through sight, from the notes of a piece of music lying before him, or from some other source, will sing, use his fingers on the upper key-board, and his feet on the lower, for the production of deep bass notes. In this case even the centres of the cerebellum are called into action.

In **Sensory Aphasia** the cerebral nervous centre is, in most cases, the first affected—possibly by embolism of the middle cerebral artery, or by other means of a mechanical nature. The subsequent softening of that portion of the brain, or hæmorrhage, is the result of the embolism. In this case the *motor* reflex speech-centre, not receiving an impulse for action, remains in repose.

Ataxic Aphasia is a loss of speech from inco-ordination of the necessary muscular movements of the throat and mouth. The muscles are not paralyzed, but their power is disturbed by the motor speech reflex centre, preventing proper co-ordination of those muscles. Nerve-fibrillæ pass from the cerebral centre of speech through the left crus cerebri and pons Varolii to the centres giving origin to the vagus, hypoglossal, trigeminus, and facial nerves in the medulla oblongata, thereby establishing its influence. In ataxic aphasia the patient knows the word wanted, but fails in articulation from disturbance of co-ordination in the muscular movements necessary. He can write, because he knows the words, but cannot read aloud what he has written. It is most frequently produced by lesions at or near certain centres of the medulla oblongata, or in an efferent nerve.

Stammering, or difficulty in pronouncing certain letters or words, is due to defective contraction of the diaphragm, causing spasmodic respiration during speech, thereby preventing air from passing through the vocal chords in a continuous flow.

Stuttering is an obstruction of speech due to spasmodic neurosis affecting co-ordination of muscles of the vocal apparatus, although respiration may be normal.

CHAPTER XXIII.

THE SKIN.

The **Skin** (or integument) subserves four purposes, chiefly, viz.: 1. It is the external covering, for the protection of the deeper parts of the body. 2. It is an excretory organ. 3. It is an absorbing organ. 4. It is the seat of the sense of touch. It consists of two principal layers—the epidermis and the dermis.

The **Epidermis** (A, B, Fig. 64), an epithelial structure on the outer surface, consists of two layers—*cuticle* (A) and *rete mucosum* (B), each of which is divided into two layers. The **Cuticle** comprises the *stratum corneum* and *stratum lucidum*; the **Rete Mucosum**, the *stratum granulosum* and *stratum Malpighii*. The **Stratum Corneum** consists of several layers of dry, horny, squamous epithelial cells, which limit superficial evaporation. The **Stratum Lucidum**, beneath, consists of several layers of cells, clear and transparent, whence it derives its name. Below that lies the **Stratum Granulosum**, the cells of which secrete a substance termed *keratin*,—a function observed in cells in all situations where the result of the secretion turns corneous. *Melanin* is the coloring matter—giving the white, brown, and yellowish color to the European, American, and Asiatic, respectively, and black to the negro. Pigment deposits, when found in the skin, as in the negro, are not formed in the epidermis, but carried there by leucocytes from the subcutaneous tissue; this is proved by the fact that white skin transplanted to a negro becomes black. The **Stratum Malpighii** covers and dips down between the papillæ and rests on a delicate basement membrane, forming the separating point between the epidermis (A, B) and the dermis (G, H).

In the superficial layers—*i. e.*, in the strata corneum and lucidum—the cells are flattened, transparent, dry, and firm, being gradually formed and incessantly renewed from beneath by proliferation of the rete Malpighii—successive layers approaching the free surface, and their contents being converted into a kind of horny matter. From the peripheral surface, the layers are constantly worn away, or desquamate in small flakes, called *dandruff*. When these two processes are accurately balanced in the phenomena of production and waste, the epidermis maintains its normal thickness; when, however, by repeated pressure, a more active cell-growth is stimulated, then the epidermis increases. This is noticeable in the hands of the laborer. The epidermis protects the sensitive and vascular cutis (true skin) from injury from without, and limits evaporation from the blood-vessels. This is proven by exposing to air the two hands of a deceased person. On one, if the epidermis be removed, the skin will become brown and dried up in the space

Fig. 64.

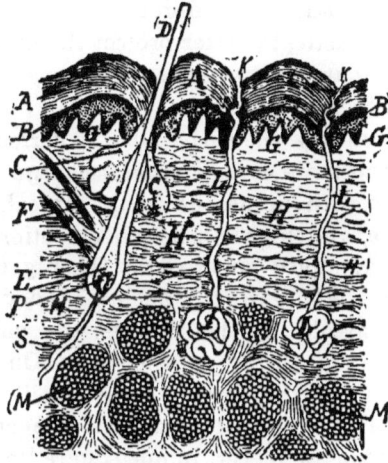

Sectional view of the skin.

A, A, Superficial layers of the epidermis, or cuticle. B, B, Deep layers of the epidermis, or cuticle. (A and B constitute the epidermis, or cuticle.) G, G, Papillæ, resting on and forming the upper part of the corium of the derma, or cutis. H, H, Deep part of the corium of the derma, or cutis. (G and H constitute the derma, or cutis.) M, M, Fat cells surrounded by fibrous tissue, situated below, and not within, the derma. C, C, Sebaceous glands. D, Hair. E, Bulb of the hair. P, Papilla at the bulb of the hair. S, Nutrient artery. F, Erector pili muscle. I, I, Sudoriparous (sudoriferous glands), or sweat glands. L, L, Ducts of the sudoriparous glands. K, K, Orifices of the ducts of the sweat-glands.

of a day; while the other, not so exposed, retains largely its natural moisture. Sunburnt complexion is due to an increase of coloring matter; which in freckles accumulates in spots. As the soft epidermic layer is transformed into cuticle the pigmentary matter disappears from its cells. In the deeper layers the cells are large, rounded, or columnar, and their contents soft and opaque. The under surface of the epidermis is accurately moulded upon the papillary layer of the derma (G), each papilla being invested, or covered, by the epidermic sheath, which also sends fine tubular prolongations into the ducts of the sudoriparous and sebaceous glands, and hair-follicles.

The **Dermis** (G, H, Fig. 64) also consists of two principal layers—the derma proper, or the superficial papillary layer (G), and the corium (H), the deeper layer. Together these layers constitute the *cutis vera*, or true skin (*Lederhaut*), which is composed of fibroareolar, yellow-elastic, and white-fibrous tissues, making it tough, flexible, and highly elastic, to better protect the internal parts from injury. These tissues are interspersed with numerous blood-vessels, lymphatics, and nerves.

The **Derma.**—Towards the surface of the derma the fibrous tissues become finer and more closely interlaced, and the most superficial layer is beset with numerous small conical, vascular, semi-transparent, flexible, and highly sensitive elevations of single and compound *papillæ* arranged in two rows (G, Fig. 64). These are essential parts in the organ of touch. They are especially numerous on the areolæ of the nipples, the palmar surface of the hand, and plantar surface of the foot. On the inner surface of the last joint of the fingers some papillæ stand in rows, others in groups, and are visible through the epidermis to the naked eye. According to Weber, in one square line of the palm of the hand there are 81 compound papillæ, with several points, and from 150 to 200 smaller papillæ, arranged in almost regular rows. At the bottom of the furrows between the ridges are minute orifices—the ducts of sweat glands. The average height of a papilla is about 1-100th of an inch, the base being about 1-250th of an inch in diameter. Some papillæ contain tactile corpuscles, others none. Those parts where tactile corpuscles in the papillæ are most numerous are the most sensitive to touch, and *vice versa*. They contain one or more nerve-fibrillæ; but those containing tactile corpuscles are constituted almost entirely of nerve-fibrillæ, while those which have no tactiles are composed mostly of blood-vessels. The axis-cylinder of a nerve-fibre enters the corpuscle and winds around spirally, while its sheath becomes identified with the outer wall of the corpuscle (Fig. 34). The tactile corpuscles are oval-shaped, and are situated in the papillæ of the derma, covered by a sheath of epithelial cells of the rete mucosum, and together are considered to constitute what is really the organ of touch (*organum tactus*). The sense of touch is most highly developed where there are the most papillæ containing tactile corpuscles. The Pacinian bodies, situated some distance beneath the derma, and not in it, are believed to be those organs by which pressure (a modification of the sense of touch) is appreciated. This implies, however, a functional relationship between these bodies, mediately or immediately, and the organ of touch (Fig. 35). The papillæ of the skin (Fig. 64) are well supplied with blood, receiving from the vascular plexus in the derma one or more minute arterial twigs, that divide into capillary loops. The capillary veins also unite in the papillæ, and pass out at their base. This abundant supply of blood in the papillæ explains their swollen appearance and reddish hue when the circulation through the skin is unusually active. The highly developed sensitiveness of a papilla depends on its tactile corpuscle. (For the different phases of the sensation of touch, see Chapter XV.)

The **Corium** constitutes the deeper layer of the skin (H, Fig. 64), and consists of a dense, strong interlacing network of oblique tensor (muscular) fibres, and various white fibrous and elastic tissues. (The latter become permanently blackened *(argyria)* in a person taking internally for a prolonged period doses of argentum nitrate.) The corium varies in thickness in different parts of the body, from a quarter of a line to a line and a half, being thickest in exposed regions, as that of the hand and sole of the foot, over the gluteal muscles and the patella, and generally thicker in the male than in the female. In the eyelids, it is very thin and delicate. The muscular membranous fibres, however, are most highly developed, and in many parts non-striated, as in the skin of the perineum, nipple, areola mammæ, and extensor surfaces generally.

The dermis, or cutis vera, is necessarily covered by the cuticle, or epidermis; for unless the very sensitive papillæ were thus protected contact with foreign substances would cause painful instead of the ordinary impressions of touch, as is proven by loss of tactile power in skin deprived of its epidermis. In parts where the cuticle is very thick, however, as on the heel, touch is imperfect or lost entirely. According to Kölliker, the thickness of the derma is from 1-8th of a line to a line and a half. The skin is thinner on the front than on the back part of the body; thinner on the inner than on the outer side of the limbs, and very thin over flexures of joints

15

and eyelids. (Concerning touch, and sensation in general, *vide* tactile corpuscles, Pacinian bodies, and chapter on Senses.)

Muscular Fibres of the Skin are voluntary or involuntary, the former, or striated fibres, being found on the head and neck only, where they radiate obliquely from the fibrous network of the derma (through which many mimic movements may be produced). The involuntary, or non-striated, muscular fibres are of three classes — muscular membranous, hair erector, and oblique tensor of the corium — and act as important regulators in the functions of the skin.

The **Arrectores Pilorum** muscles (F, Fig. 64; and *m*, Fig. 65) originate in the derma, and are partly attached to the hair-follicles. They also send down loop-shaped attachments to the elastic framework of the corium. During a slight contraction of these muscles, they relieve the sebaceous glands of their contents, and have no effect on the hair-follicles; their severe irritation, however, causes traction on the hair-follicles, resulting in erection of the hair (noticeable on the back of an angry dog, or cat).

The oblique tensor muscular fibres, with the elastic tissue, constitute, mainly, the corium. The elastic tissue exerts compression, and prevents the lymph-spaces from overfilling with the fluid of broken-down tissue, in which action the oblique tensors assist the compressing force, in response to the action of delicate nerve stimulus. The muscular and elastic framework establishes an inhibitory apparatus, distributing pressure, traction, and relaxation equably, and influencing the movements of the fluids, nutrition, and the interchange of gases.

Tension of the Skin.—Temperature exerts an influence on the muscular apparatus of the skin. A medium temperature causes moderate tension; a higher temperature, entire relaxation. In the senile skin, especially on the face and neck, the muscular fibres change and degenerate, causing the skin to wrinkle.

The **Subcutaneous Tissue** beneath the corium (M, Fig. 64) is composed of connective tissue, fat and fat-cells filling up the spaces, acting as a soft elastic pad, and having very little conducting power; hence, it regulates to some extent the radiation of heat. (*Vide* Fat). It is blended, by strong interlacing fibrous bands, with the corium.

The skin is provided with two kinds of secreting glands, *sudoriparous* (or sudoriferous), and *sebaceous;* and is furnished with two kinds of appendages, *hair* and *nails.*

The **Sudoriparous** or **Sweat glands** (each a small coiled tube, or convoluted reddish body) are the organs by which a large portion of aqueous and gaseous materials are excreted, and are found in almost every part of the skin (I, Fig. 64). On the palm of the hand and sole of the foot, their orifices occur in regular lines between the ridges of the papillæ; on other parts of the body they are scattered irregularly. The efferent duct (L), on leaving the gland, passes upwards in a spiral or cork-screw manner as far as the papillæ of the corium, and lies between them. From there it passes straight to the cuticle, when it again resumes a spiral course and opens on the surface of the cuticle by a slightly oblique, valve-like aperture (K). The diameter of the duct (L) is about 1-1700th of an inch. The glands are soft, of a reddish color, and more or less flattened by the pressure of one on the other. They are imbedded in connective tissue, and permeated by a network of capillaries. Their average size is about 1-50th of a line, their number over two millions; their situation is in small pits in the deep parts of the corium, or below in the subcutaneous adipose tissue (I). The palm of the hand contains about 2800 orifices in each square inch. Krause calculated that the orifices of the sweat glands of the entire body represent an opening equal to about eight square inches of space. Those of the axilla (odoriferous glands of Horner) are among the largest, and secrete a peculiar odorous matter. Moll's glands open into the hair-follicles of the eye-lashes. The peculiar, bitter, yellow, adhesive semifluid substance secreted by cells of ceruminous glands in the external auditory meatus is called *cerumen* (earwax). Its function is the protection of the meatus and the membrana tympani from dust and insects. These glands are of a brownish color, and are simple modifications of (not differing much in structure from) the ordinary sudoriparous glands.

Sweat is not the product of the sudoriparous glands only. The intercellular passages of the epidermic cells open into the sweat-pores (ducts), by which the tissue fluid of the epidermis escapes externally. Consequently, sweat contains substances not secreted by sudoriparous glands (analogous to urine, which contains ingredients not formed by Malpighian bodies alone). The dog has sudoriparous or coil-glands, but does not sweat, its sweat-pores opening into the hair-follicles. Sweat is usually formed so gradually that its watery portion escapes by slow evaporation, termed *insensible perspiration*. When it becomes profuse and collects in drops, it is called *sensible perspiration*. Sensible perspiration is the most direct means of lowering the temperature of the body. Increased sweating is usually accompanied by dilatation of the cutaneous vessels; and, like other secretions, such as saliva, is under the reflex influence of a special nervous centre in the medulla oblongata, acting through its nerve-fibres upon the cells of glands. Sweating, therefore, occurs not only when the skin is red, but also when pale. It occurs, besides, when the cutaneous circulation is languid, as when accompanying syncope, or immediately preceding death.

On account of the presence of secreting glands, the orifices of which open externally, the skin becomes an excreting organ. The excreted fluid (sweat) consists of urea, extractive matter, carbonic acid gas, and water, the latter amounting to about two pounds in 24 hours. The sebaceous secretion (sebum) is of a more oily nature, and is designed to keep the skin moist and pliable, and prevent a too rapid evaporation of moisture from the body. If, in a warm-blooded animal, these secretory processes are stopped — *e. g.*, by varnishing the skin all over with an impermeable substance — the animal will die.

The **Sebaceous Glands** secrete the **Sebum,** a suet-like fluid (C, Fig. 64). They are organs of a whitish hue, found in nearly all parts of the substance of the corium or subdermoid tissue, except on the palm of the hand or sole of the foot; they are most numerous in the integument of the scalp, face, nose, armpits, and anus. The largest glands of this class are the *Meibomian,* found in the eyelids. The tubes (ducts) of sebaceous glands are very small. They are filled with an opaque, white substance. One or more open generally into the hair-follicles, though sometimes on the surface, as on the labia minora, prepuce (Tyson's glands), and red margins of the lips of the mouth.

Secretions of the sebaceous glands, hair-follicles, and epidermic intercellular matters passing into hair-follicles, become so intimately intermingled as to form a single homogeneous product. It consists of oily and extractive matter, caproic and butyric acids, and cast-off epithelial cells, the former of which gives it its peculiar odor. The purpose of the sebaceous secretion is to keep the skin moist, to moderate evaporation, and to guard it from the effects of the long-continued action of moisture. The secreted product is similar to that on the body of the *fœtus in utero.* It must, therefore, be reckoned among the excretions. What part it has in the process of purifying the blood is unknown. Sebaceous glands often contain a parasite termed *acarus folliculorum.* Sebaceous matter, when first secreted, is fluid in consistency; but in the ducts often hardens, as on the nose, where it can be squeezed out in a wormlike form—the so-called *comedo*.

The **Appendages of the Skin**—the *hair* and *nails*—are modifications of the epidermis.

Hair (D, Fig. 64; and A, Fig. 65).—The skin (except that of the outer surface of the eyelids, palms of the hands and dorsal surfaces of the third phalanges, and soles of the feet) is covered with hair. The hair may be fine and delicate, or coarse and strong. The individual hairs are situated in depressions of the skin termed hair-follicles (G), and extend obliquely into the deepest layer of the dermis, or into the subcutaneous tissue, and have at the bottom

Fig. 65.

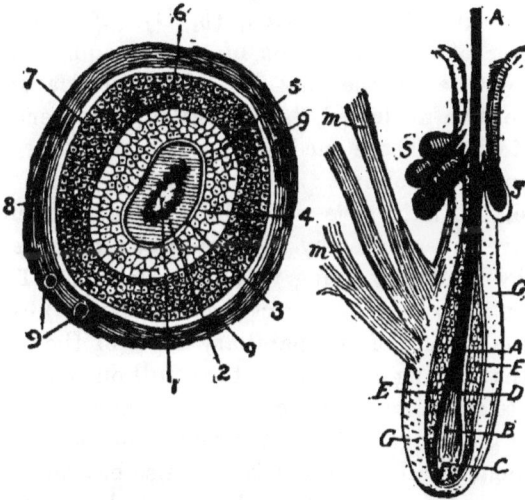

Diagram of transverse and longitudinal sections of a hair in hair-follicle.

1, Medulla, or pith of the hair. 2, Cortex, or cuticle layer of the hair. 3, Hyaline membrane between the hair and follicle. 4, Huxley's layer of internal lining of the hair-follicle. 5, Henle's layer of internal lining of the hair-follicle. 6, Middle layer of the follicle. 7, Fibrous layer between the middle and external layers. 8, External layer of the follicle, with blood-vessels. 9, Blood-vessels.

A, Hair. B, Bulb. C, Papilla. D, Hyaline membrane and internal sheath of the follicle. E, Middle layer of sheath of the follicle. G, External layer of sheath of the follicle. S, Sebaceous glands. m,m, Erector pili muscles.

small papillæ (C), from the surface of which the hairs take their rise. The wall of the follicle is lined continuously with squamous epithelial cells (4, 5, 6, Fig. 65), resting on a layer of the fibrous connective tissue (8 and G) in contact with the surrounding corium. The follicle appears simply as a deep depression in the surface of the integument. The wall of the follicle is generally channeled by one or more ducts of sebaceous glands (S), some of which discharge into the hair-follicle, and others on the free surface of the skin.

The hair grows from a papilla (C), at the bottom of the follicle, the papilla supplying the material for its nourishment. It is lined with a layer of squamous cells. This lining makes a turn upward at the side of the papilla, so that the horny cell layer of the hair and follicle face each other. Their union is prevented by a layer of fusiform cells which interposes a transparent, glossy, hyaline membrane between, (3, D). For reasons stated, the outer layer of hair, consisting of everted squamous scales, is termed the cuticle of the hair (2). This cuticle surrounds a single layer of non-nucleated, flat, elongated, homogeneous cells, in the center of which is the medullary space, sometimes filled with pith, but generally hollow.

The root of the hair rests upon the papilla, the cells of which form successive layers whereby the hair grows towards the surface of the skin by gradually rising within the follicle. The hair and follicle lay obliquely, but the erector pili fibres (m) extend angularly from the surface of the papillary layer of the dermis to the follicles, so that during contraction they pull on the follicles, causing the hair to stand erect on the skin. When a hair is broken off by pulling, it grows again; but if the root below be extracted (bulb B), the epithelial wall along the follicle also becomes injured, and the hair will not be reproduced. Hair is shed in certain animals periodically by being lifted off from the papilla through the contractive, or pinching, force of the dermis on the hair-follicles. As soon as the pressure from below upward is lessened the new hair commences to grow, pushing the old outward, the old hair falling off when the new reaches the surface. In the human scalp, in old age, the fibres of the dermis gradually contract, become hard, and produce pressure on the hair follicles; in consequence thereof the hair-stem becomes thinner, assuming the *lanugo* (woolly) form; or, if the contraction and pressure be complete, the hair papillæ atrophy, the follicles permanently collapse, and the hair growth is destroyed. Again, during a severe illness, the old papillæ may dry up and the hair become loose. When new papillæ form new hair grows, gradually pushing the old out of the follicles.

Gray Hair.—When the pigment formation in the cortex becomes defective, the hair turns gray, as in old age; or if the medulla, or pith, become defective, the center of the hair becomes filled with air, giving it a silvery appearance.

Blood-vessels, Nerves, and **Lymphatics.**—The hair-follicles, sudoriparous and sebaceous glands, are surrounded by areolæ, or spaces filled with adipose tissue. These spaces are the channels through which the vessels and nerves are distributed to the more superficial strata of the corium and papillary layer. The dermis is supplied with numerous lymphatics. The blood-vessels form two systems, superficial and deep. The superficial supply the papillary layer; the deep, the fatty tissues, hair-follicles, sebaceous and sudoriparous glands.

A **Hæmorrhage** in the skin does not rise above its level. It may take place in one of three ways:—1. Rupture of vessels from external violence or internally by increased blood-pressure, resulting in effusion into the integument. 2. In vascular degeneration in disease of the walls of vessels. 3. By transudation of blood through the vascular walls. Bluish-red, greenish-yellow, and yellowish-brown spots, are produced by the progressive involution of the stagnated blood.

The loss of water through the skin in a healthy adult in 24 hours by sensible perspiration, adding the carbonic acid gas, salts, and other excretory matter, is about one and a half to two pounds; so that in this time there may be a loss of 1-28th of the weight of the body by urine, respiration, and perspiration.

The toughness, resistance, flexibility, and elasticity of the skin admirably fit it for protecting the organs beneath. The horny layer of the epidermis, hair, and subcutaneous fat serve for the retention of the natural warmth of the body. The skin is also an organ of absorption of substances in solution sufficiently dilute to penetrate the horny layer of the epidermis. They are taken up by the cells and ducts of glands, absorbed by capillary veins and lymphatics, and pass into the general circulation, acting in the same manner as if taken internally. Water and fluid or semifluid forms of nourishment, especially milk, may be administered through the skin; also many medicines, as by inunction or in solution, except strong astringents, which are difficult or impossible of absorption. While a strong solution of an alkali, especially soda, hardens the epidermic cells, a weak solution softens and swells them.

The **Nails** on the fingers and toes consist of horny, homogeneous cells resting on a basement layer, and run posteriorly and laterally in a groove. The posterior white part, termed *lunula*, comprises the nail-roots. The growth of the nail is forward from the *matrix*, or posterior aspect, extending from the extreme posterior portion of the roots to the anterior border of the lunula. The cells of the matrix secrete the nail substance, which deposits in layers. The new secretion gradually pushes the previously formed part forward. The thickness of the nail-body is the same from the lunula to the anterior projecting edge. The posterior groove in which the nail-roots are lodged prevents growth backward.

CHAPTER XXIV.

THE KIDNEYS.

The **Kidneys** are two glandular organs situated in the abdominal cavity, one on each side of the vertebral column. They are held in position by the entering and departing vessels and by a quantity of surrounding fat, somewhat firm. The anterior portion of the right kidney lies close to the liver, and to the descending part of the duodenum and ascending colon; while the anterior portion of the left is in close relation with the spleen, pancreas, stomach, and descending colon; only the surrounding fat separating it from these organs. The kidney is a smooth, dark-red, compressed oval body, with a hilum (notch) on the inner surface in connection with an anterior *sinus* (cavity), where the renal vessels, nerves, and excretory duct have their entrance and exit (5, 6, and 7, Fig. 66). The ordinary size of a kidney in the adult is about four inches in length, two in width, and one in thickness. It weighs about five ounces. Its external portion, termed the *cortical substance* (1), is about one-fourth of an inch thick, and constitutes about three-fourths of the gland. It consists of a multitude of glomerules (Malpighian bodies), nerve-fibres, capillary blood-vessels, and lymphatics, in connection with the uriniferous tubules. Between these organs is interposed a small quantity of very fine areolar tissue. The glomerules give it a granular appearance. The central portion of the kidney is the *medullary substance*, and consists principally of straight uriniferous tubules, arranged in a series of various sized cones forming the *renal pyramids* (2), from ten to fifteen in number and extending from the papillæ (3) through the medullary substance, to near the surface of the cortical structure. The apex of the pyramid-shaped arrangement of uriniferous tubules forms a nipple-shaped point at the pelvis, and is termed a *renal papilla* (3). The uriniferous tubes are the channels through which the urine flows from the cortical substance (1) into the pelvis (4)—*i. e.*, the funnel-shaped upper portion of the excretory duct, or ureter (5). Both the pelvis and the ureter are lined with mucous membrane, containing transitional epithelial cells, sometimes intermixed with columnar epithelial cells. The cavity of the pelvis is subdivided into four or five portions, called *calyces* (small cup-like membranous canals). Into each calyx three or four renal papillæ project, so that the urine dripping from the uriniferous tubules, and through the papillæ, is received by the calyces, conveyed to the pelvis, and descends through the ureter (5) into the bladder. In this manner the calyces, pelvis, and ureter, together, form the excretory duct for the urine.

The kidneys are well supplied with blood-vessels, nerves, and lymphatics, held in place by connective tissue. The renal artery (6), is short and thick, proceeds directly from the descending aorta, and supplies the arterial blood; which, having thus considerable force, easily penetrates the kidney and communicates with the

Fig. 66.

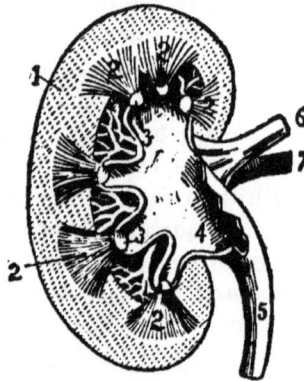

Diagram of the kidney.

1, Cortical substance. 2, Renal pyramid. 3, Renal papilla.
4, Renal pelvis. 5, Excretory duct (ureter). 6, Renal artery.
7, Renal vein.

glomerules, uriniferous tubules, and veins. The capillaries of most of the arterial and venous branches ramify in the cortical substance, forming the glomerules.

A **Glomerule** (Malpighian body, or corpuscle), is a dense cluster, or round, ball-like mass, formed by capillary arteries, veins, and the endings of nerve-filaments, enclosed by a capsule (Bowman's capsule, A and B, Fig. 67), about 1-144th to 1-80th of an inch in diameter. This capsule is the commencement of the uriniferous tubule (E). The capsule and tubule consist of basement membrane, lined with glandular epithelial cells. The tubules are convoluted, and unite as they pass along to form the pyramids (2, Fig. 66), which at the cortical portion may consist of hundreds of tubules, while at the opening of a papilla (3) they are few in number.

The glomerules, or Malpighian bodies, cause the cortical substance to appear granular. Into each glomerule, arterial and venous branches enter (1, C and D, Fig. 67). The function of the glomerule is to separate the superfluous water from the blood. This

Fig. 67.

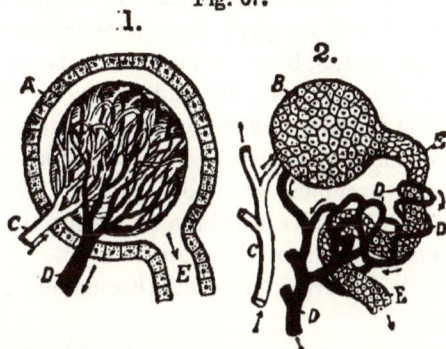

Diagram of a glomerule, or Malpighian body, of the kidney.

1, Longitudinal section of a glomerule. 2, The glomerule intact, being completely surrounded by its capsule. A and B, Bowman's capsules. C, C, Branch of the renal artery. D, D, Branch of the renal vein. E, 1, Longitudinal section of a uriniferous tubule. D and E, 2, Indicating the network of a uriniferous tubule and its capillary blood-vessels.

is, possibly, simply a dialytical, or straining process, forcing watery parts of the blood from the glomerule into its surrounding cavity. The capsules (A and B) and the uriniferous tubule (E, E) are lined with glandular epithelial cells, which absorb, separate, and secrete from the blood various elements forming urea, uric acid, creatine, and the inorganic salts, such as chloride of sodium and phosphate of soda. The cells perform the functions of absorption, separation, and secretion, and transmit the saline elements, in solution, to the interior cavity of the capsule and uriniferous tubule (E, Fig. 67), where the aqueous portion coming from the glomerule unites with the saline solution and passes into the pelvis as urine. The arrangement of the convoluted portion of the tubules (D, E) in the cortical substance is such that it forms a network similar to that of capillary blood-vessels; so that within the cortical substance the network of both are so closely interwoven that the secreting

epithelial cells absorb or separate the salts from the blood with facility. The tubules of the network of the cortex are winding in
their course—*i. e.*, they descend and partially return many times,
forming loops (2, D and E), called Henle's loop-tubes. On nearing
the papillæ the tubes discharge into one another, their number becoming smaller in the formation of the pyramids, finally passing
in a straight line to the papillæ.

The function of the kidneys is to separate certain matters from
the blood that would be injurious to the system if retained. The
urine, therefore, for the most part, consists of waste matters. Its
fluid and solid constituents bear no constant ratio to each other,
the proportion of fluid present depending upon the amount of
fluid taken with the ingesta, and on arterial capillary fullness and
pressure; while the amount of solids contained depends on the kind
of food taken and previous waste of tissues. The Malpighian
bodies and uriniferous tubules are the most essential organs in the
kidneys for the performance of their function.

Lymphatic capillaries are distributed throughout the parenchyma of the kidneys.

The nerve-fibres are derived from the renal plexus. They carry
impulses for sensory, vaso-motor, vaso-dilator, and for secretory
functions.

The **Urine** is a clear, amber-colored fluid, normally of an acid
reaction; average specific gravity, 1020. Its composition is:—

Water...		950.00
Nitrogenous organic substances.	Urea...................................	26.20
	Creatine..............................	0.87
	Sodium and potassium urates.........	1.45
	Sodium and potassium hippurates....	0.70
	Mucus and coloring matter...........	0.35
Mineral salts.	Sodium biphosphate..................	0.40
	Sodium and potassium phosphates...	3.35
	Lime and magnesium phosphates.....	0.83
	Sodium and potassium chlorides.....	12.55
	Sodium and potassium sulphates.....	3.30

1000.00 ,

The quantity of urine secreted daily is about 1400 cubic centimetres (about 48 ounces). It is discharged from four to six times
during that period. Should the secretion of urine become completely
suppressed, the elements of urea remaining in the blood would soon
exert a poisonous effect, especially on the nervous system. This
usually, and speedily, results fatally. If at one time of the day
urine has a specific gravity of 1012, and at another 1028, the

variation may be accounted for by the different kinds of food taken, and may be considered normal; but either 1012 or 1028, if long continued, would be an indication of abnormal conditions.

Urea constitutes the characteristic and most important solid ingredient of urine. It crystallizes in white, glistening, streaked, four-sided prisms; or, when very rapidly crystallized, it forms in small, white, silky needles (Fig. 68). Its taste is bitter, like that of salt-petre. Urea is an *amide of carbonic acid;* for urea, when added to water and heated to 200° in a tightly closed tube, breaks up into carbonic acid and ammonia. Generally, some of the mucus of the

Fig. 68.

DIAGRAM OF APPEARANCE OF UREA WHEN CRYSTALLIZED.

bladder is voided with the urine and acts as a ferment; so that in a few days, at ordinary temperature, urine is rendered ammoniacal. The normal stimulus to micturate is the sensation excited by the action of the urine on the mucous membrane of the bladder. An impulse is forwarded through afferent (sensory) nerve-fibres in communication with the hypogastric plexus to the nervous centre of micturition in the spinal cord at the fourth lumbar vertebra. This centre reflexes the impulse by sending an efferent (motor) impulse through nerve-fibres in communication with the sacral plexus to the sphincter vesicæ, which dilates, thus allowing the urine to flow out through the urethra. The power of the will exerted on this centre excites a similar reflex motor impulse to micturate; or the power of the will may restrain the reflex action, though for a limited time only. The average amount of urine excreted by an adult in twenty-four hours contains about 33 grammes (500 grains) of urea; and from about seven to ten grains of uric acid. An artificial urea may be obtained by the action of ammonia on ethyl carbonate; or by the action of ammonia on carbon oxydichloride; also by the action of heat on ammonia carbamate and carbonate. Urea forms approximately one-half of the entire quantity of the solid (saline) ingredients of urine. The quantity of urea in the urine depends upon the rapidity of tissue decomposition, especially nitrogenous, and the kind of food taken. It is increased by the consumption of animal food and by increased muscular activity; but decreased by a vegetable or non-nitrogenous food diet, such as starch, sugar, and fat. The materials entering into the formation of urea are, therefore, in the tissues and in the blood; but urea, as such, does not normally exist therein, being formed completely only by the secreting cells of the capsules (Bowman's) enveloping the Malpighian bodies, uriniferous tubules, and the ducts of the sebaceous glands. In a pathological condition, when the Malpighian bodies and tubules of the kidneys, or the ducts of the skin, become obliterated, the urea already formed in these organs becomes absorbed by the lymphatics, and may be carried into all fluids and parts of the body. Urea is a secreted excrementitious compound, which has to be speedily removed, or its accumulation in the system exerts fatally paralyzing effects, especially on the nervous centres.

From 50 to 60 grammes of solids are discharged daily with the urine by the healthy adult. When organic salts, such as acetates, lactates, and tartrates, are taken, they are changed in the blood into carbonates, and the urine becomes alkaline. It is claimed that about 10 to 18 grammes of sodium acetate, taken internally, causes alkalinity of the urine within twenty minutes. The sodium and potassium urates are also increased (like urea) by nitrogenous food, the result of alkaline bases with nitrogenous mineral acid. Any substance that will neutralize an acid is called a *base*, and a compound of an acid and a base is termed a *salt*. In the urine some of the alkaline sodium phosphate is replaced by acid sodium biphosphate, which gives urine the property of reddening blue litmus paper, although urine contains no free acid.

Uric acid is combined with sodium and ammonium urates, the sodium predominating. Uric acid does not exist in the urine in a free state, but in combination, principally as sodium urate, and is believed to combine with some of the sodium, the base of sodium phosphate. The remainder of this salt being converted into acid sodium bisulphate, the urine thus acquires an acid reaction. As almost any acid added to urine decomposes the urates, uric acid is set free and crystallizes (Fig. 69). By using largely for a time fruit and vegetables containing tartrates and citrates of the alkaline bases, the urine becomes alkaline from the presence of alkaline carbonates, since the conversion of the tartrates and citrates into carbonates takes place in the blood. Continued mental labor increases the alkaline phosphates, while the earthy phosphates are increased by **nervous** diseases, the urine, having an acid reaction, **holding** the earthy phosphates, such as phosphates of lime **and** magnesium, in solution.

Fig. 69.

Diagram of uric acid crystals.

The **Bladder** is well supplied with nerve-fibres, blood-vessels, and lymphatics. The nerves are branches from the sacral plexus of the spine and hypogastric plexus of the sympathetic.

Incontinence of Urine is often due to diminished sensitiveness in the mucous membrane of the bladder or sphincter vesicæ, and the urine is evacuated unconsciously.

The **Retention of Urine** in old people is generally due to the hardening of the *prostate gland*, which presses upon the urethra. The efferent (motor) reflex impulses then have diminished power to dilate the sphincter vesicæ.

The general opinion is "that the kidneys in the early period of fœtal life are very small and imperfect, their functions being performed by the **Wolffian bodies,** two organs analogous to the kidneys in structure. These afterwards become atrophied and then disappear. The kidneys grow rapidly, making their first appearance just behind the Wolffian bodies, and take their place at the end of the second month of fœtal life." It seems more likely, however, that the Wolffian bodies noticed in early fœtal life develop into the suprarenal capsules. That these bodies ever performed the functions of the kidneys is rather problematical, as it has not been proven that the kidneys are active, any more than are the lungs or the intestines of the fœtus.

Disease of one or both kidneys may affect the body to such an extent that excrementitious matters accumulate in the blood, so contaminating it as to render it unfit to carry on the vital processes on which the activity of the brain depends. This results in delirium or unconsciousness.

During health the cells lining the uriniferous tubules contain very small oil-globules. In the pathological state called **Bright's disease** they are more numerous and very much enlarged, causing pressure on the neighboring capillary veins and lymphatics, and obstructing the circulation. The reason of the great pressure on the capillary arteries is that the blood flows direct from the abdominal aorta through the short and thick renal artery; and, meeting with obstruction in the veins and lymphatics, the *albumen* in the serum of the blood exudes with the water and salts, and so gets into the urine.

In acute inflammation of the kidneys (*nephritis*) there is a deep-seated pain in the back, on one or both sides, aggravated by pressure or sudden change of position. The urine becomes scanty, highly colored, albuminous, or bloody; and, if allowed to stand, deposits pus and sediment. The patient is more or less feverish, generally troubled with rigors, nausea, vomiting, and constipation. In chronic inflammation the symptoms are milder. In both acute and chronic inflammation, the blood may become contaminated, as in Bright's disease, from the want of proper purification by the kidneys, when many secondary affections may arise.

The **Suprarenal Capsule** is a small, flattened, glandular body of a yellowish color, weighing from one to two drachms, and belongs to the class of ductless glands. It consists of a cortical and medullary substance, with numerous blood-vessels and nerves, and is located over the anterior and upper portion of each kidney. Dr. Addison, of Guy's hospital, London, first pointed out that a certain disease of the suprarenal capsule or capsules is associated with a deposition of pigment in the skin, which assumes a deep bronze color. Addison's disease is sometimes termed *suprarenal melasma*, or bronze-skin disease, and is characterized by general debility, loss of muscular power, remarkable weakness of the heart's action, extreme prostration, breathlessness on the slightest exertion, dimness of sight, functional weakness and irritability of the stomach, a peculiar and uniform discoloration of the skin of a brownish olive-green hue—like that of a *mulatto*. The morbid changes in the capsules are of a soft deposition, degenerating into a yellowish-white opaque matter, followed by abscess, or drying-up into a chalky mass. The only probable aid is nourishing food and tonics to keep up the general health. The progress of the disease is slow, extending over a period of from one to about five years. Then death follows by heart failure, as if its natural stimulus, the blood, had ceased to act.

CHAPTER XXV.

THE MUSCULAR TISSUES.

The fully developed muscular tissues consist of two distinct kinds of textures: 1, *Striped fibres;* 2, *non-striped fibres.* Muscular tissue composed of striped fibres is called *striated muscle;* that composed of non-striped fibres is called *unstriated muscle.* Respecting the function of these two kinds of muscle-tissue, no strict classification can be made. In general, however, it may be said that the striated muscles are those that perform functions pertaining to *animal* life, that they are under the control of the will (except the muscular structure of the cardiac organ), and are capable of rapid movement involving the whole muscular organ; while the unstriated muscular tissues perform functions pertaining to organic (vegetative) life, are involuntary in their action, contract slowly, and principally affect the tissues in their own extent only.

In the embryo, up to the third month, this difference of the fibres in the muscular tissue is not apparent, though it must exist in a rudimentary way; for, under normal conditions, tissues do not change in organic structure. It is questionable if the change of cartilage into bone can be regarded as a fundamental one. There is an increase in the quantity of tissue merely — the parts increasing in size and weight, though the increase does not involve a change in elementary composition. During development, the number of elementary fibres is not increased from that which existed in the muscle tissue of the fœtus. Muscle tissue forms about 45 per cent of the entire mass of the adult body.

The **Striated Muscle** consists of the following structures:—
1. The **Perimysium Externum**, a connective-tissue envelope (fascia, aponeurosis) surrounding the muscle (1, Fig. 70). It sends partitions (septa) inward, supporting the nerves, lymphatics and blood-vessels, and then forming the perimysium internum. 2. The **Perimysium Internum** (2), a delicate connective-tissue envelope derived from the perimysium externum. It separates the muscular fasciculi, or bundles of muscular fibres (3), and sends partitions inward, which constitute the *endomysium* (4).

16

Thè **Fasciculi** (3) of a small muscle are as long as the muscle itself, while those of long, large muscles are from one to two inches in length. 3. The **Endomysium** (4) consists of delicate connective-tissue, derived from the perimysium internum, and forms par-

Fig. 70.

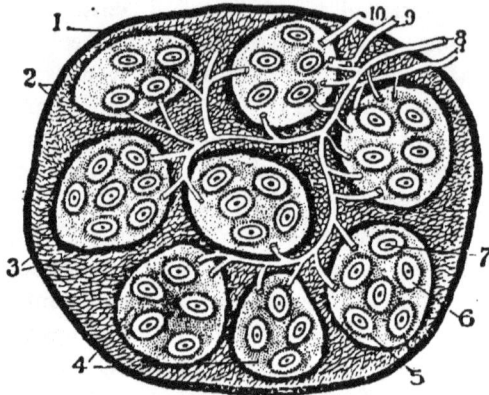

Diagram of transverse section of a striated muscle.

1, Perimysium externum. 2, Perimysium internum. 3, Fasciculus. 4, Endomysium. 5, Sarcolemma, surrounding the muscular fibres. 6, Muscular fibre. 7, Muscular fibre-cell. 8, Blood-vessels. 9, Nerve. 10, Lymphatic vessel.

titions or septa extending inward and forms the sarcolemmæ. The endomysium and sarcolemmæ separate the muscular fibres, they carry and support the capillary vessels and nerve-fibres within the fasciculi of voluntary or striated muscles. These muscles are composed of a great number of fasciculi, each consisting of many muscular fibres running in parallel lines, and held together in bundles *(fasciculi)* by the perimysium internum. The fibres of a fasciculus are not all of equal length. Their ends are generally attached to adjacent fibres by the ensheathing **sarcolemma** (5), a strong, adhesive substance, which continues from each fibre as a tendinous fibrilla to the termination of the fasciculus, and thence to the end of the muscle, where the sarcolemmæ and endomysium blend together and merge with the perimysium externum, forming the **tendon,** which is attached to the periosteum of a bone. It is noticed that the larger nerves and vessels of a muscle are situated between the perimysium externum and perimysium internum (8, Fig. 70), while the nerve-fibrillæ (9) and capillary-vessel

networks (10) run in the endomysium; *i. e.*, externally to the sarcolemma and muscular fibres, but within the perimysium internum.
The motor nerve-fibres enter the muscular fibres (Fig. 37), but the
capillary vessels do not. 4. The **Sarcolemma** (5) is a delicate,
structureless, transparent, and elastic envelope, surrounding the muscular fibre, and holding it in proper shape. It sends transverse
partitions (the membranes of Krause) across the muscular fibre
(2, at C, Fig. 71) at regular intervals, separating each space (1),
which contains a nucleated cell (5). These cells not only maintain

Fig. 71.

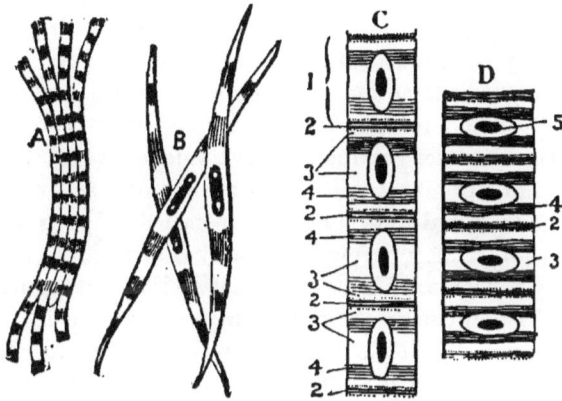

Diagram of muscular fibres.

A, Four partly isolated striped fibres. B, Three isolated unstriped
fibres, each having a nucleated cell. C, One striped fibre elongated
(in a state of repose), and its nucleated cells. D, The same striped
fibre (C) contracted—shortened and thickened—as during activity.

1, Disk, or division, of a striped fibre as it appears under polarized
light. 2, A faint layer (Krause's membrane), constituting the septum,
or stria, between the disks. 3, A single refractive substance (Hensen's
disk). 4, Double refractive substance. 5, Nucleus within the cell.

the function of nutrition and reparation of the muscular fibre, but
also during activity secrete from the blood (in a way similar to that
of glandular cells) the contractile substance (*myosin*) for the striæ
and sarcolemma. The contractile substance is dispersed more or
less throughout the muscular fibre, but the greater quantity is held
in the sarcolemma and membranes of Krause (*striæ*). Myosin is a
nitrogenized substance forming the chief constituent of muscles.

The **Muscular Fibre** consists of a long, thin cylindrical mass, covered by a sheath termed *sarcolemma*. It contains, besides nucleated cells (C, D, Fig. 71), a sarcous substance. These fibres vary in size with the age of the person, averaging in the adult from 1-500th to 1-250th of an inch in diameter, and from one-fourth to one-half of an inch in length. In very short muscles, such as the stapedius in the middle ear, the fibres are as long as the muscle. The average distance between Krause's membranes (*striæ*) is 1-9400th of an inch, the extremes being 1-15000th to 1-6000th, according to the extent of the contraction or relaxation of the fibre.

Physiologically, the properties of muscles are fourfold: (1) Irritability; (2) Elasticity; (3) Tonicity; (4) Contractility.

Irritability of a muscle is that inherent property by which its cells are brought into action through an impulse received from a nerve, from the blood, from a stimulus applied directly to them, from excrementitious matter retained in its tissue, or from temperature. Every muscular fibre is supplied by a branch of a *motor* nerve-fibre, ending in the terminal plate (Fig. 37) so-called; that is to say, one original nerve-fibre may, through its many ramifications (2, Fig. 33), supply many muscular fibres. The nerve-fibre enters the muscle generally at a point near its middle, at a place the least affected by its contraction. Of all the tissues of the body, the superficial muscles and the skin are most abundantly supplied with nerves. They run along the outside of the sarcolemma of each muscular fibre, similar to the blood-vessels; but a motor nerve-fibre ultimately penetrates the sarcolemma, its axis-cylinder leaving the nerve fibre-sheath and the white substance of Schwann, and terminating in the muscular fibre in a special expansion called the terminal plate (Fig. 37), or motorial end plate. The *sensory* nerve-fibres terminate, not in the muscular fibres, but between them, in the endomysium and sarcolemma. It is a peculiar fact that the power of an impulse conveyed through a *motor* nerve-fibre, producing contraction of a muscle, is largely determined by the length of the nerve-fibre. It seems to be a general law of animal life that the strength of an impulse through a motor nerve-fibre increases correspondingly with the length of the fibre. Hence, nervous actions on the muscles of the hands and feet are more powerful than on muscles (comparative size considered) nearer nervous centres. A further peculiarity of striated muscles is, that the more frequently they are called into action, if that action be not too prolonged, the larger and more vigorous they become, as is evinced in all forms of judicious exercise. This follows from the increased demand for oxygenated blood and appropriate nourishment. The number and size of tendinous fibres in connection with the muscular fibres also play an important part in the power of muscular action. An abnormal temperature of the body affects the property of irritability; it is either much increased or much lowered. Any interference with the reflexes to muscles, or with their nutrition, enfeebles or wholly destroys their irritability. Nothing so effectually impairs their nutrition as disuse, and nothing so completely destroys it as the division of the nerve (destroying transmission of reflexes)—involving also the loss of contractile power. If called into action, however, by massage, or other stimulating treatment during the period of the muscle's impairment, and regular nutrition be continued, its irritability is retained.

Muscular tissue is excitable independently of nervous influence; there may, therefore, be loss of muscular excitability without loss of nervous excitability, and *vice versa*. Curare paralyzes the motor nerve endings without affecting the contractile power of the muscle. In curare poisoning, reflex impulses through the motor nerve-fibres have no effect on the muscle; still the muscle retains its excitability to stimulation applied directly to the muscle. Landois states "that ammonia, carbolic acid, or lime water applied to muscles produces movement, but not when applied to the nerves."

Elasticity of a muscle is that inherent property by which its fibres and tissues are enabled to stretch and elongate, or shorten, to a certain extent, without injury. It moderates the effects of powerful, sudden stimulation, acting gradually, and preventing detachment at the points of origin and insertion. This is an important endowment, well illustrated in the extension and flexion of limbs. Normally, all fibres do not act simultaneously; for when some cease to contract, their extremities are acted on by the contractions of adjacent fibres.

Tonicity of a muscle is that inherent quality of sensitiveness by which its fibres and tissues become aroused to functional activity in response to either direct or indirect stimulation. In health the tonicity of groups of muscles is so adjusted as to constitute a mutual counterbalance. This counterbalance may be destroyed, however, by paralysis or impaired nutrition. The tonicity of one set of muscles is then weakened; as in lead-palsy, for instance, in which the extensors of the forearm and hand lose their natural strength, so that the contraction of the flexors keeps the fingers constantly bent upon the palm. In health the tonicity of the flexors usually exceeds that of the extensors — in profound sleep, for

example, when the influence of the will and the nervous system
is lulled, the hands being partially closed.

Contractility of a muscle is that inherent, chara cteristic property by which its fibres and tissues forcibly contract and shorten
in a particular direction in response to the will, external irritation,
or stimulation; exhibiting itself in dual form, *contraction* and *relaxation*. It is most impressively shown in the voluntary muscles and
heart, which, when in action, exhibit powerful con tractions, alternating with relaxations. *Striped fibres*, as previously pointed out, torm
the greater mass of the quick-acting muscles, which are generally
attached to bones by tendons. They are commonly controlled by
the will, and are, therefore, called *voluntary muscles;* though not.
altogether correctly so, as many of them—*e. g.*, the heart; and along
the spine, neck, and middle ear—are beyond the will's control.
The sphincter muscles and diaphragm are composed of striped
fibres. The enclosing structure of cavities is supplied with longitudinal, circular, and oblique muscular fibres, which act simultaneously in contraction; *e. g.*, those of the heart, blood-vessels,
lymphatics, stomach, gall-bladder, intestines, uterus, Fallopian
tubes, urinary bladder and ureters. In all cases where muscular contraction takes place, some other tissue is moved, or tends to move.

The contractile movement of the striped fibre is wonderfully
rapid, occurring at the instant of stimulation. The motor nerves
of the muscular system are derived from the cerebro-spinal system.
The power of contractility is made possible, principally, by the presence of the nitrogenized compound **myosin,** a semifluid substance
contained in the striœ of the muscular fibres and in the sarcolemma
surrounding these fibres. Muscular contractions occur alternately
in the fasciculi, passing along the fibres within the fasciculi from
the point of stimulation, in a wave traveling from 6 to 16 feet
per second. The velocity corresponds to the state of vital activity,
and of temperature. When vital activity is very low, stimulation
causes only a local contraction and slight elevation at the point of
contact. Stimulation to muscular contraction takes place in one
of two ways: (1) By the application of the stimulus to motor nerves,
termed indirect stimulation; (2) to the muscular tissue itself, termed direct stimulation. When muscles are active, the vessels dilate,
more blood and nutrition are supplied, and more carbonic acid elements are formed and gas given off at the organs of exit. If excessively stimulated for a time the cells of muscles become exhausted
(as in glandular secreting cells); loss, therefore, of contractile

substance follows. Muscles are extremely vascular. The capillary arteries, veins, and lymphatics run within the endomysium (4, Fig. 70). They are distributed in the endomysium (outside of the sarcolemma), in lines nearly parallel with the muscular fibres, but do not enter the fibres. The nutriment (*pabulum*) required for the maintenance and growth of these fibres is drawn by absorption through endosmotic imbibition from the network of capillary vessels, through the sarcolemma. The excrementitious substances are drawn by the absorbent vessels through exosmotic absorption from the muscular fibres, too, through the sarcolemma. The power of imbibition of nutrient matter and absorption of excretory matter is increased during contractile activity. The supply of blood is not only required for the nutrition of the muscular tissue, but also affords the oxygenous condition requisite for its action. That blood may exercise its salutary influence it must be arterialized. The muscles soon lose their contractile power if the arterial blood supply has been suspended, either by cessation of circulation or want of proper aeration.

The activity of the muscular structure (striped fibres) of the heart, being necessarily constant, *reparation* must take place *pari passu* with its waste; during health no sense of fatigue is experienced in the heart. In other striated muscles, under the control of the will, reparation is not so rapid, and their prolonged exertion induces impairment and fatigue, attributable to defective nutrition. In this case prolonged rest is rendered essential to recovery. Sodium chloride and the potassium salts are especially necessary for the normal nutrition of muscles. If these salts are withheld for some time, the muscles atrophy, and the central nervous system and the digestive apparatus become disturbed. Of all saline solutions, chloride of sodium is the most necessary; again, of all the salts, it injures the nervous and muscle tissues the least.

The muscular tissue, *i. e.*, its contractile substance, tires out sooner than its nerve terminals. In paralysis of the muscle, its nutrition becomes defective and the *myosin* measurably disappears; the muscle loses its contractile power correspondingly. By massage the flow of blood and lymph is facilitated, nutrition favored, and waste products in fatigued and weakened muscles are removed. During muscular fatigue, an increase of decomposition of matter in the muscle takes place. By local absorption of nutrition, or by passing through the blood of the muscle a weak solution (6 to 8 per cent) of chloride or carbonate of sodium, or a constant electrical

current, the muscle soon recovers its loss. Generally, under artificial stimulus, muscular contractility continues after death for a time,
if the muscle be kept cool and moist; but is lost as soon as the
myosin coagulates or degenerates. Striped muscular fibres retain
their contractility much longer than unstriped, the latter losing it
almost immediately after death.

Muscular Action is referable to *muscular contraction*, involving a shortening and hardening of the fibres; and the bringing of
their ends nearer together (D, Fig.71). The fibres, including the
sarcolemma, possess the property of contraction—not the fat, neither
the connective tissue, nor endomysium of the muscle. In regard to
simplicity of mechanism and amount of work performed, muscles
are vastly more perfect than any human device; the more they are
normally exercised the stronger they become and the more work
they can accomplish. The more fibres contained and distributed
transversely in a muscle the stronger it is. In its formation some fat
or adipose tissue is interspersed more or less within the fibres, between the fasciculi and fibres, and between the fasciculi themselves.

Muscle-tissue when at repose is neutral or slightly alkaline. The
contractile activity moves the muscle and the part to which it is
attached. This causes partial degeneration, and at the same time
involves the mortality of certain constituents of the muscular fibres
and connective tissue; then acid and excrementitious substances
form more quickly in the muscle. Vital tissue acid (a form of lactic
acid) is the only acid formed directly, and is the result of what is
known as the *katabolism* of muscular tissue. The acidity is neutralized by the alkalinity of the blood, resulting in carbonic acid
elements, which are carried away from the muscular tissues by the
efferent vessels to the organs of exit. The quantity of carbonic acid
elements evolved is far greater during physical activity than during
the process of muscular nutrition simply. It is removed from the
tissues principally by the capillary veins, while other excrementitious substances are removed by the lymphatics. How this activity
is effected, may be understood when we consider that *protoplasm* is
a substance subject to incessant molecular changes. The changes
in the vital economy termed *metabolism*— going on under the influence of the activity of the vital principle, guiding the chemical combinations within the vital animal economy — are properly
divided into two kinds: One, termed *anabolism*, implies an upward
series of chemical combinations, by which the latent energies of inert
food are transformed into the living energy of the protoplasm of

the bodily tissues. The other, termed *katabolism*, involves a down-ward series of changes in the vital economy, by which the living tissues are partly broken up and the waste matter is set free. As protoplasm itself undergoes changes according to the *pabulum* sup-plied, so it modifies the tissues accordingly. The form and function of protoplasm, therefore, are the effects of certain causes. Again, living matter (protoplasm) acts upon, and is powerfully acted upon by, the matter surrounding it. The food entering the body holds energy in latent form (chemically speaking), and supplies the material for physical action. By manifold transformations and chemical changes, the food materials are changed into true tissue *pabulum*, first exhibiting energy as heat only. It finally becomes a part of the living cell-substance—*i. e.*, *protoplasm*—and is then enabled to give forth energy that may be utilized in nervous and muscular action. It is through the process of nutrition that the various parts maintain the same general conditions of form, size, and composition that they attained by growth and development. In health an adult maintains for a number of years nearly the same features, weight, and size, although simultaneously the tissues continually undergo decay and renovation. Prolonged mental exer-tion is often followed by an increased amount of alkaline phos-phates, found in the urine, arising from the increased destruction of the cell-tissues of the brain; yet the brain substance does not diminish, owing to the active reparative processes at work. The same is true of the muscular cell-tissue; an increased amount of exercise is directly followed by an increased excretion of the ordi-nary products of the decomposition of cells and of nitrogenous tissues, assimilation of pabulum at the same time restoring the wasted tissue.

Muscular activity increases heat, extended muscular contractions evolving more heat than limited ones. When we ascend a flight of stairs two steps at a time more heat and fatigue ensue than when we ascend the same flight leisurely, one step at a time. The trans-formation of physical energy within the body takes place chiefly in the muscular tissues and in the secreting glands.

The **Unstriped** or **Involuntary Muscular Fibres** con-sist of a series of longitudinal or cylindriform filaments, somewhat granular, pale, and flattened, from about 1-300th to 1-100th of an inch long, and from about 1-3000th to 1-1500th of an inch in thick-ness, and pointed at their extremities (B, Fig. 71). These fibres are usually marked at intervals by one or two minute elongated cells containing nuclei with nucleoli. The cells are from about 1-500th to 1-200th of an inch long, and from 1-5000th to 1-2500th of an inch broad. Nerve-fibrillæ are in connection with these cells, and run longitudinally with the unstriped fibres. The impulse through the nerve-fibre to the cell induces the contraction of the sarcolemma. As the unstriped muscular fibres are differently constructed from the striped, their stimulation, contraction and time of action differ from them also. During contraction the muscular fibres become shorter and thicker. The tonicity of unstriped fibres exerts a slow, prolonged contractile movement; more so than that of striped fibres. Generally, unstriped fibres, unlike the striped, have no fixed points of attachment; but form continuous investments around cavities within the body, as in the intestinal canal, bladder, uterus, and blood-vessels; or are dispersed through the substance of tissues, as in the cutis of the skin, to which they impart a contractile property. In the walls of capillary vessels these muscular fibres lie in single layers, and are, as it were, consolidated; the end of one being joined to the body of the other. In other situations, as in the walls of the intestines, stomach, or bladder, they exist in double and triple longitudinal, oblique, and transverse layers. Unstriped fibres contract very slowly, requiring a longer time than the striped for the nerve influence to take effect, so that an interval exists between the instant of stimulation and that of contraction. Such muscular fibres are generally found in internal organs, where gradual and lasting contractions are required. Their nerve supply emanates generally from the sympathetic system, and they are, therefore, uncontrolled by the will. Those of the alimentary canal are prompted to action more readily by direct alimental stimulus than in any other way. Weak solutions of mineral salts, bile, glycerine, and salts of iron, also stimulate them. The striped fibres of muscles, on the other hand, are more readily called into action by stimulus through the nerves of the cerebro-spinal system.

CHAPTER XXVI.

REPRODUCTION.

Reproduction of animal life is accomplished by one of three processes—fission, gemmation, or fecundation.

Fission (dividing or splitting) is the reproduction of an organism by its division into two parts, each part possessing potentially all the functional powers of the parent.

Gemmation, or budding, is where an organism originates by a bud arising from some part of the parent structure, the pedicle or stem of the bud gradually disappearing, when, being liberated, the new organism assumes a perfected form, closely resembling that of the parent.

Fecundation is reproduction resulting from the impregnation of an egg (ovum), requiring in its accomplishment two individuals of the same species but of opposite sex—the male furnishing the sperm-cell (spermatozoon), and the female the ovum, containing the germ-cell.

Propagation by fission and gemmation occurs in well-marked forms among the lower orders of animal life only. Amœba and some varieties of infusoria present good illustrations of reproduction by division (or fission); other varieties of infusoria, and polyps, are examples of reproduction by gemmation.

Reproduction distinguishes in a striking manner organized living beings from unorganized, inert bodies. Reproductive cells go through a series of developmental changes, which do not interfere materially with the general life of the parent organism. If the fertilized ovum of any species be deposited where the special conditions favoring its development are present it acquires maturity; otherwise not.

The **Uterus,** or womb, of the human female is a pear-shaped organ (Fig. 72; and *a.* Fig. 73), situated in the pelvis between the bladder, and the rectum. It is divided into three parts, a fundus, body, and neck (cervix); the neck being about half as long as the body, but narrower. The uterus is composed principally of three kinds of tissue layers; externally, of a serous (5, Fig. 72); internally, of a mucous (7); and intermediately, of a thick fibro-muscular coat (6 and 9).

Up to about the fourteenth year, or time of puberty, the uterus is small. At this time, however, it increases in weight to about an ounce and a-half, and in length to about two and a half inches.

Fig. 72.

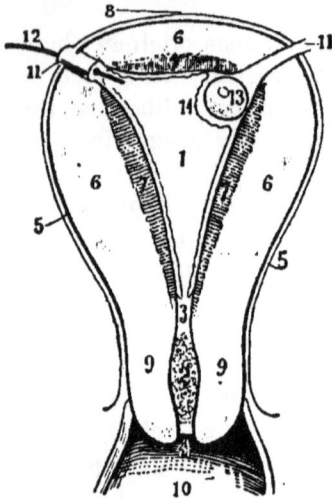

Diagram of the uterus.

1, Cavity of the uterus. 2, Cavity of the neck (cervix) of the uterus. 3, Os internum. 4, Os externum, or mouth of the womb. 5, 5, Serous layer. 6, 6, 6, Fibro-muscular layer. 7, 7, 7, Mucous layer. 8, Fundus of the uterus. 9, 9, Neck of the uterus. 10, Vagina. 11, 11, Fallopian tubes. 12, Bristle run through the channel of one Fallopian tube. 13, Fertilized ovum. 14, Membrana decidua.

Its internal cavity is constricted just below the midd'e, forming the *os internum* (3), this marking the dividing line between the cervix and the body. The orifice (4) at the lower part of the cervix is termed the *os externum,* or mouth of the womb. The cervix projects partly into the vagina (10). The dome (8), or upper part of the uterus, is called the *fundus.* At its junction with the body the *Fallopian tubes* (11,11) enter. The *ovarian* and *round* ligaments (*p, h,* Fig. 73) lie imbedded in the *broad ligament* (*e*), which is a sero-membranous structure on each side, serving to hold the uterus in position. The muscular wall of the uterus is about half an inch in thickness, the walls being separated internally from side to side by a triangular cavity (1, Fig. 72). In antero-posterior section, however, this cavity appears as a narrow slit lined with mucous membrane (7), which constitutes nearly one-fourth of the entire thickness of the uterine wall and is firmly adherent to its muscular structure (6). Whenever the uterus contracts normally its mucous membrane is thrown into numerous folds. The membrane is pale-red in color, and still paler, and much thinner, in the cervix. At the os externum the membrane is very thin, especially in the virgin. Its surface is lined with columnar epithelial cells, which during pregnancy become changed to the pavement variety. It also contains numerous orifices of the so-called *glandular follicles of Naboth,* the length of which is proportionate to the thickness of the membrane. Their walls are lined with ovoid epithelial cells. The follicles furnish an increased secretory surface, and their function is to secrete an alkaline substance. They sometimes close up and become swollen with their secretion, and the swelling may be mistaken for ova or a fœtus. Such enlarged follicles are termed *ovula of Naboth*, or *hydatids*. They appear as yellowish vesicles of various sizes, especially in the mucous coat of the cervix uteri (2, Fig.72) where the *glandulæ folliculi Nabothi*, about 10,000 in number, lubricate the parts, during deliv-

Fig. 73.

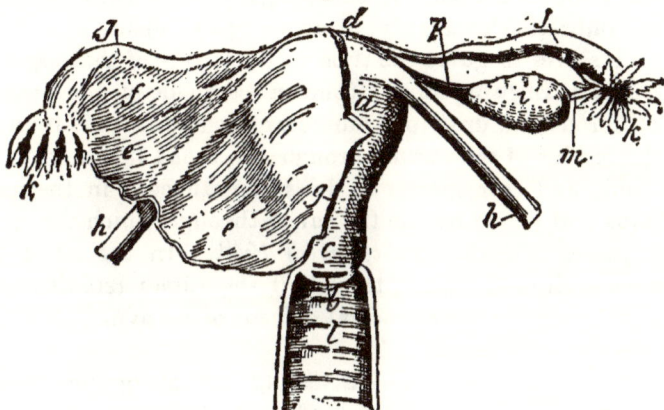

Diagram of the uterus and its appendages.

a, Uterus. b, Os uteri. c, Neck of the uterus. d, Fundus of the uterus. e, e, Broad ligament. f, Ovary within the broad ligament. g, Indicating the margin where one-half of the broad ligament is cut off. h, h, Round ligaments. i, Ovary, the broad ligament being removed. j, j, Fallopian tubes, or oviducts. k, k, Fimbriated extremities of the Fallopian tubes. l, Vagina. m, Fringe-tube connecting the ovary with the Fallopian tube, or oviduct. p, Ovarian ligament.

ery, with an alkaline secretion. In morbid conditions, this secretion is often increased and of acid reaction. The entire uterus is well supplied with lymphatic and capillary blood vessels, very distinct around the follicular orifices, especially when the uterus is excited to functional activity, as during the menstrual period.

The **Fallopian Tube,** or **Oviduct** (one on each side of the uterus, 11, Fig. 72; and *j*, Fig. 73), through which the ovum passes from the ovary (*i*) to the uterus, is situated at the upper part of the broad ligament. Its canal at the uterine end is so small as to barely admit the passing of a bristle through it (12, Fig. 72). It enlarges towards its outer extremity, which is trumpet-shaped, and from its fringed appearance is termed the *fimbriated extremity* (*k*, Fig. 73). Wilson says: "The remarkable manner in which this circular fimbriated extremity applies itself to the surface of the ovary during sexual excitement, has gained for it the additional title of *morsus diaboli.*" Each Fallopian tube is about four inches long, and is lined internally with mucous membrane and ciliated epithelial cells, the direction of the vibrations of the ciliæ being towards the cavity of the uterus. One of its fringed processes (*m*) forms a connecting passage-way leading from the outer end of the ovary (*i*) to the Fallopian tube. A short ligamentous cord proceeds from the fimbriated extremity, attached to the distal end of the ovary, and serves to guide the tube in its seizure of that organ.

The **Round Ligament** (*h*) (one on each side of the uterus), formed of fibrous and unstriped muscular tissues, extends from the side of the fundus of the uterus through the broad ligament, to the labia majora, and corresponds to the spermatic cord in the male. In the fœtus and young female the canal through which the round ligament passes is called the *canal of Nuck*. In the adult this canal is generally obliterated, though it sometimes remains open, affording a chance for the abnormal descent of an ovum, or it may become the seat of a hernia.

The great serous membrane investing the abdominal viscera passes over the upper part of the bladder and downward between it and the uterus, to a point midway between the *os externum* and the *os internum*, thence upward (5, Fig. 72) over the fundus of the uterus, then downward (5) behind the vaginal fornix where its dip forms the *pouch of Douglas*. Thus the peritoneum forms a cap, as it were, embracing the whole of the superior portion of the uterus antero-posteriorly. From the sides of the uterus to the inguinal region the peritoneum lies doubled, forming the broad ligaments.

The **Broad Ligament** (*e, e*, Fig. 73), as previously stated, consists of two layers of the peritoneum, which encloses the Fallopian tube, ovary, and ovarian and round ligaments, besides holding in its meshes numerous muscular fibres. These fibres, by their contraction, exercise an important physiological function in bringing all the structures into harmonious action. The broad ligaments hold the uterus in position in the center of the pelvic cavity, and admit of considerable antero-posterior movement of the uterus, to accommodate the distension of the bladder and the rectum.

Blood Supply.—The two uterine arteries are derived from the internal iliacs, and pass to the sides of the neck of the uterus.

They are very tortuous in their course, admitting of great uterine distension during pregnancy without diminution of calibre. They supply blood to the uterus and give off branches that pass between the layers of the broad ligament to the bladder. The two ovarian (spermatic) arteries arise from the front of the descending aorta just below the renal arteries. They, too, pass between the folds of the broad ligament, supplying the ovaries and Fallopian tubes, and a small branch anastomoses with the uterine artery. The veins accompany the arteries.

Nerves.—The uterine nerves are derived from the inferior hypogastric plexus situated at the side of the rectum, vagina, and bladder. The plexus are partly formed by fibres from the second, third, and fourth sacral nerves. The nerves accompany the uterine arteries, supplying the uterus and Fallopian tubes. The ovarian nerves emanate from the spermatic plexus, and supply the ovaries and the broad and round ligaments.

The **Lymphatics** are large and numerous, especially in the impregnated uterus, and communicate with the pelvic and lumbar lymphatic glands. Thrombosis of these vessels, especially during child-birth, interferes with the removal of decomposed matter, giving rise thereby to puerperal fever.

The **Ovary** (*i*, Fig. 73) in the female is analogous to the testicle in the male. The former produces the germ-cell; the latter, the sperm-cell, or spermatozoon, for future offspring. The ovary (one on each side of the uterus) is imbedded in the broad ligament posteriorly, and has a special white fibrous tissue covering, termed *tunica albuginea ovarii* (1, Fig. 74). The inner extremity of the ovary is connected with the uterus by the ovarian ligament (*p*, Fig. 73); its outer, with the fimbriated process (*m*) of the Fallopian tube. In appearance the ovary is white, smooth, and plump, about one and a-half inches long, three-fourths of an inch wide, and half an inch thick. Its weight is from one to two drachms. It consists of unstriped muscular fibres, interlaced with connective tissue, blood-vessels (8, Fig. 74), lymphatics, nerves (9), and, scattered throughout its mass, from 15,000 to 25,000 small, round, transparent sacs of various sizes and stages of development termed *Graafian vesicles*, to which some apply the term *ovisacs*, imbedded in the ovarian stroma and containing a clear, colorless, albuminous fluid. Each vesicle also contains an ovum, or *vesicle of Baer* (6, 6). The *membrana granulosa* lies between the Graafian vesicle (2) and the zona pellucida (3), and consists of nucleated cells. Each Graafian vesicle has a minute follicular duct terminating at the surface of the ovary. The majority of ova in the ovary remain so minutely small (7, Fig. 74) that a female of eighteen may have from 30,000 to 50,000 in both ovaries. Before puberty a large ovum is about 1-500th of an inch in diameter, and is to all appearances a cell; its zona pellucida, or vitelline membrane (3), being its wall; the vitellus (5), its cell contents; the germinal vesicle (4), its nucleus; and the germinal spot, its nucleolus. The diameter of a mature ovum (2 to 5), is about 1-120th of an inch; its germinal vesicle (egg-cell, nucleus, or vesicle of Purkinje), about 1-720th of an inch, surrounded by a delicate membrane; and the germinal spot, which is about 1-4000th of an inch.

Fig. 74.

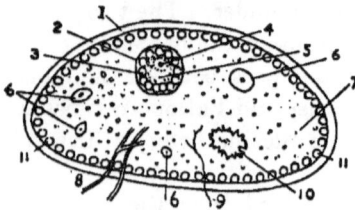

Diagram of an ovary.

1, Tunica albuginea ovarii. 2, Graafian vesicle, or ovisac, surrounding the ovum. 3, Zona pellucida, or vitelline membrane. 4, Germinal vesicle containing the germinal spot. 5, Yelk of the ovum. 6, 6, Ova gradually enlarging to maturity similar to the one already mature (3). 7, Minute ova (undeveloped). 8, Blood-vessels. 9, Nerves. 10, Corpus luteum, or ruptured spot of a Graafian vesicle, after liberating the ovum. 11, Cells lining the internal surface of the tunica albuginea ovarii from which the ova originate.

Ovulation.—From an early period in fœtal life, when the ova form in the ovary, they grow by increase of the yelk surrounding the germinal vesicle (4). Each ovum is finally inclosed by a membrane termed the *zona pellucida* (3), situated within a small sac termed *Graafian vesicle* (2). The maturation and discharge of an ovum takes place in the human female, generally every twenty eight days, and is normally associated with menstruation, and constitutes the phenomenon of *ovulation*. Not only the Graafian vesicle, but all parts of the internal generative system participate in this periodical manifestation. The enlargement of the Graafian vesicle, with its contained fluid and ovum, is due to increased blood vascularity, which goes on until the vesicle ruptures. This generally taking place at the menstrual periods, though during sexual excitement the *morsus diaboli* may prematurely induce a rupture and liberate an ovum from the ovary. At the time the ovum escapes from the ripe Graafian vesicle the germinal vesicle and germinal spot move towards the periphery of the yelk nearest the point of rupture, the germinal spot placing itself in a position favorable to a visit from the spermatozoon. The bursting of the Graafian vesicle takes place at the tunica albuginea (1, Fig. 74), so that the ovum is received by the oviduct (*m*, Fig. 73), and passes through the Fallopian tube into the uterus. Previous to the time of rupture, the internal layer (3, Fig. 74) of the Graafian vesicle becomes yellow and waxy by the aggregation of minute oil-globules on its surface. After the rupture the yellow appearance becomes more distinct by direct exposure of the layer (3); hence the name *corpus luteum* (10) applied to the laceration or scar resulting. This name applies generally to all corpora lutea, whether impregnation has taken place or not. The following distinction, however, is observed: When impregnation of the ovum does not take place the congestion of the uterus, ovaries, and neighboring parts disappears, the place of rupture of the Graafian vesicle shrivels, contracts, heals rapidly (within two or three weeks), becomes absorbed, and the spot of rupture and its yellowness is scarcely discernible; it is then termed *false corpus luteum*. If, however, pregnancy occurs, then the functional activity of the uterus excites an increased afflux of blood to the ovaries and neighboring parts as well as to itself; hence, they remain more or less congested. The healing of the rupture is thus prevented, the edges of the laceration degenerate, and the ruptured spot becomes larger, and is supposed not to heal until after delivery. Such a scar on the ovary is termed the *true corpus luteum* (10, Fig. 74). It is believed to be the subsequent course which determines the corpus luteum of pregnancy—it is more fully developed and of longer duration, harder, larger, and at a later stage has a distinct white covering with a large stellated cicatrix. Congestion of the ovary and neighboring parts may continue, however, for some time after a catamenial ovulation, without impregnation, the healing being prevented and the scar remaining large. Hence, the existence of a large corpus luteum on the ovary affords no evidence whatever of intercourse having taken place. On the other hand, its absence would not warrant the belief that conception had never taken place. The terms *true* and *false* corpora lutea are, therefore, not strictly definite, and may lead to serious mistakes in a medicolegal sense. The discovery of a fertilized ovum (embryo), and its development in the female generative organ, is perhaps the only positive proof of conception. (*Vide* Moles and Hydatids.)

17

The **Testicle** (T, Fig. 75) is a glandular structure, consisting of about 400 lobules—*lobuli testis* (C). These lobules are composed of minute convoluted tubes, lined with cells termed *vesicles of evolution*, forming groups of four to six; and from the nuclei of these vesicles, or cells, the spermatozoa are developed, constituting the active generative element of semen. The tunica albuginea testis (A) sends out fibrous septa between the lobuli testis, thereby separate them.

Spermatozoa (S, Fig. 75).—At and after the age of puberty the testes of the male secrete a fluid termed *semen*, which is a glutinous, whitish substance, with a peculiar odor, containing minute elongated bodies, the spermatozoa (S), which are thread-like bodies, each about 50 micromillimetres long, with a pear-shaped head (1), rod-like middle part (2), and ciliated caudal prolongation (3). During sexual excitement some of the cells within the lobuli testis (C) burst, setting the spermatozoa free into the seminal fluid secreted by the organs F, G, H, I, Fig. 75, and discharged by the vas deferens (J) through the male organ of coition into the vagina of the female. The spermatozoa are capable of movements upward in the vagina and

Fig. 75.

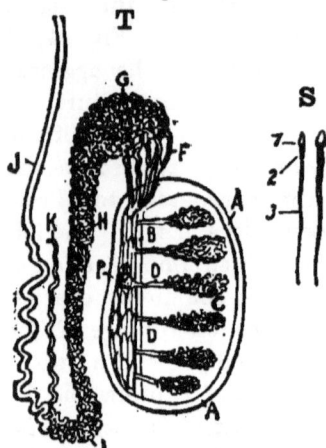

Diagram of the testicle and its ducts (**T**), and two spermatozoa (**S**).

A, A, Tunica albuginea testis. B, Mediastinum testis. C, Lobuli testis. D, D, Vasa recta. F, Vasa efferentia. G, Coni vasculosi, constituting the globus major epididymis. H, Body of the epididymis. I, Globus minor epididymis. J, Vas deferens. K, Vas aberrans. P, Rete testis, forming a layer interspersed with capillaries of the spermatic artery between B and the internal portion of the tunica albuginea.

uterus — no doubt due not only to their own propelling power, but also favored by the peristaltic action of muscular fibres in the mucous membrane of the vagina, uterus, and Fallopian tubes. The extent of their movement is from 0.05 to 0.5 of a micromillimetre per second during several days. When spermatozoa are absent from seminal fluid, as in debility, or old age, impregnation is impossible. It is their absence from the semen that causes hybrids to be sterile.

Fecundation is the fertilization of the ovum by the spermatozoa of the semen. Up to the third month of gestation the ovum is termed the *embryo*, after that the *fœtus*. At the time of sexual excitement the erectile tissue of the male organ is acted upon reflexly through the nervous centre situated in the lower portion of the spinal cord, culminating in orgasm and erection of the muscular fibres of the male sexual organ, through which the semen is directly ejaculated by the contraction of the fibres of the vas deferens (J, Fig. 75). The contraction of the fibro-muscular tissue surrounding the prostate gland and the urethra at the neck of the bladder pre-

Fig. 76.

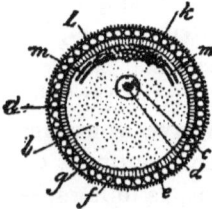

Diagram of an impregnated
ovum.

a, Zona pellucida, or vitelline membrane. b, Yelk. c, Germinal vesicle. d, Germinal spot. e, Space left by retraction of the yelk from the vitelline membrane. f, Layer of cells that remained from the tunica granulosa which encircled the ovum in the Graafian vesicle; (a and f, greatly enlarged, form the vitelline membrane. k, Germ-mass. m, m, Germinal membrane, a portion only being indicated.

vents regurgitation towards the bladder. Thus, during the act of coition, the semen passes into the vagina at or near the os uteri externum. The discharge of semen into or near the os is not absolutely necessary to impregnation, however, as there are cases on record where the seminal fluid came in contact with the pudendum simply, the hymen being intact; yet the spermatozoa made their way inward to the ovum and fertilization followed.

Impregnation.—The meeting place of the spermatozoon and ovum is generally about the middle of the Fallopian tube (*j*, Fig. 73). As a rule, impregnation occurs within a week after menstruation, but it may take place earlier or later (*Vide morsus diaboli*). The zona pellucida, or covering (3, Fig. 74; and a, Fig. 76) of the ovum, is a delicate structure. Some believe that it has several fissures, through which the caudal portion of the spermatozoon enters and thus fecundates the ovum. It has been surmized also (as white blood-corpuscles, under certain circumstances, penetrate the walls of blood-vessels and may be transformed into pus cells) that in the case of impregnation it be possible that the caudal end of the spermatozoon actually effects the entrance through the delicate zona pellucida of an

unfissured ovum.   As soon as impregnated, the germinal vesicle of
the ovum changes into an embryonic cell, while yet in the Fallopian tube, and the yelk becomes cloudy and indistinguishable.
The first step in embryonic development is the segmentation or
cleavage of the ovum, though still enveloped in its zona pellucida.
The transformations in the germ-cell are so rapid that by the time
the ovum arrives at the uterine cavity the entire yelk is changed
into a germ-mass, which becomes flattened against a segment of the
internal surface of the zona pellucida, as indicated at *k*, Fig. 76.
The aggregation of the germ-mass forms the *area germinativa*, which
soon exhibits three different classes of cells, each class forming a
separate layer; the three layers constituting a single membrane.
Now, this membranous formation (as indicated at *m, m*, Fig. 76) is
termed the *blastoderm*, or *germinal membrane*.   Its external pellicle
(ectoderm) is the *serous* layer; the internal layer (entoderm), the
mucous; and the intermediate, or middle layer (mesoderm), the
vascular.   The entire embryonic structure now continues to develop
from this blastoderm, or germinal membrane, as follows: —

THE BLASTODERM, OR GERMINAL MEMBRANE.

From this membrane originate the Ectoderm, Mesoderm, and
Entoderm.

The **Ectoderm** (ectoblast, or epiblast) is the external germ-
layer, çell-wall, or yelk-membrane.   From it are derived the nervous system, dermal and epidermal tissues, voluntary motor apparatus (including muscles and bones), eyes, ears, nose, mouth, and
anus.   The mouth and anus are formed by depressions into the
entoderm.

The **Mesoderm** (mesoblast) is the middle germ-layer.   From
it are derived the involuntary motor apparatus, alimentary canal,
heart, lungs, lymphatics and blood-vessels, vascular glands, secretory, excretory, and generative organs.

The **Entoderm** (endoderm, or hypoblast) is the internal germ-
layer.   From this layer are derived the epithelial lining of the alimentary canal, air-passages, and ducts of secreting glands, and the
endothelium of the walls of serous cavities.

These three principal layers are intimately connected, so that
cells of each class of tissues remain associated with others, more or
less, to keep up during growth the connection necessary between
the great nervous centres and the tissues of the body.   It is believed that during the early days of the embryo the cells of the central

portions of these layers gradually separate by rapid growth, yet retaining sufficient the intimate relation of the different apparatus and systems, such as the nervous, muscular, vascular, and other tissues of the body.

The ovum is lodged at the mucous membrane in the uterine cavity (13, Fig. 72), the most extraordinary activity of its cells soon attracting surrounding epithelial and other cells of the uterine mucous membrane, when the *membrana decidua* is developed.

The **Decidua** (14, Fig. 72) originates from the mucous membrane of the uterus, which becomes thickened, highly vascular, and softened, forming the special envelope (*decidua*) for the impregnated ovum. The part of this envelope lying on the muscular wall of the uterus is termed the *decidua vera*. The folds of the decidua which grow over (next to) the ovum, are termed the *decidua reflexa*. The part of the decidua vera which takes part in the formation of the placenta is termed the *decidua serotina*. The decidua does not grow nor extend to the external mouth (only to the internal os) of the uterus, as the neck (2) becomes closed by a plug of mucus.

The **Chorion** is a thin, transparent membrane, surrounding the fœtus and amnion. It is formed from the vitelline membrane of the ovum, and becomes covered with villi. It is composed of two layers—an external vascular serous layer, or *false amnion*, and an internal fibrous layer termed *allantois*—which project from the middle and internal germinal layer, and is continuous with the intestinal cavity of the embryo. Afterwards, out of the lower portion of the allantois is formed the bladder cf the fœtus, and its upper layer furnishes a vascular membrane, which, with some of the large villi of the chorion, penetrate the decidua serotina, uniting it and the chorion firmly together, forming thereby the **Placenta.** Thus the combined maternal and fœtal circulation is established, the villous portion remaining with the fœtus and the decidua serotina with the uterus of the mother (7, Fig. 78). The placenta is of threefold importance—it is the organ, or channel, through which the circulation, nutrition, and excretion, of the fœtus are accomplished. At birth, by the ligation of the umbilical cord, the hæmoglobin in the ligated vessels is converted into bilirubin, producing jaundice in the new-born infant for the first few days.

Extra-uterine Pregnancy.—Normally, the uterus is the receptacle in which the new individual develops. The embryo may, however, form in the Fallopian tube (j, Fig. 73) or in the fringe (m) near the ovary in the broad ligament. Such cases are termed extra-uterine fœtation. That such abnormal forms of pregnancy are of rare occurrence may be thus explained: The spermatozoon traveling through the uterus and Fallopian tube, and not coming in contact with the ovum, arrives at length in a labyrinth—*i, e.*, in the fimbriated extremity of the Fallopian tube (*k*, Fig. 73)—and is lodged in the wrong fringe, perishes, and shares the fate of excreted mucus. There are a thousand chances to one that this will be its fate. However, should it accidentally find its way into the fringe-tube connected with the ovary, and here fertilization be consummated, then the relaxed fringes stop the further progress of the ovum, and extra-uterine pregnancy is established.

The fimbriated extremity of the Fallopian tube may be regarded as serving more than one purpose; *viz.*, to grasp the ovary (morsus diaboli) and extract an ovum during sexual excitement, and to furnish a trap for the spermatozoa so as to prevent it finding the ovum near the ovary, thereby preventing extra-uterine pregnancy.

Graviditas in Substantia Uteri takes place where there are one or more fissures, or crevices, within the mucous membrane of the uterus, the fertilized ovum accidentally becoming lodged in the depth of such a fissure.

Twin, Triple, or Quadruple Pregnancies generally occur when more than one ovum is liberated from the ovary or ovaries, or when spermatozoa enter both Fallopian tubes and meet an ovum in each at the same time, or in one tube after one act of coition and in the other after a subsequent. At an average, in the white race twin pregnancy occurs once in about eighty times; triple, once in 6000; quadruple, once in about 800,000 times.

The **Mamma**, or breast (A, A, Fig. 77), is a rounded eminence on each side of the anterior aspect of the thorax, between the sternum and the axilla. It may be said that they are glands accessory to the generative system. Each mamma is provided with from fifteen to twenty excretory tubes termed *tubuli lactiferi* (F), terminating at the nipple (B) by minute orifices. Around the nipple is a colored circle, or *areola*, which becomes darker during the period of pregnancy, and remains darker in women who have borne children than in those who have not. The substance of the mamma consists of minute lobules (*alveoli*) containing the milk-secreting cells, and are clustered together in racemose or conglomerate bunches, forming large lobes (D). These are situated within (C), which forms their envelope, composed of areolar and connective tissues. The ducts of (D) communicate with and discharge into dilatations (F) of the lactiferous ducts. The secreting lobes and their ducts are interspersed with areolar and fibrous tissue. The entire substance of the mammary gland is supplied with blood from the thoracic branches of the axillary, the intercostals, and internal mammary arteries. It is also well supplied with lymphatics and nerves. The nerves are branches from the anterior and lateral intercostal cutaneous. The secreting alveoli and the internal aspect of the walls of the lactiferous tubules are lined with epithelial cells. The capillary arteries furnish the blood for the nutrition of the gland, as well as material to the cells for the secretion of the milk. The elaboration of milk takes place in the secreting cells in the walls of the alveoli. The milk globules (fat of milk) are formed in the protoplasm of the epithelial cells. These milk (or fat) globules are much larger at the begining of lactation; such milk is called *colostrum*. In the dilatations (storehouses) (F, F, F), the milk accumulates during the time the child is not nursing.

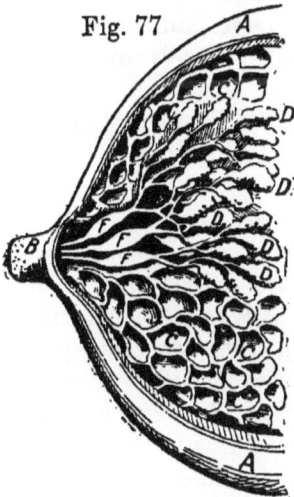

Fig. 77

Diagram of a section of the mammary gland.

A, A, Section of the skin on the surface of the gland. B, Nipple with minute orifices of the milk ducts. C, C, C, Areolar and connective tissues forming roundish structures containing the nerve-fibrillæ, lymphatics, and capillary blood-vessels; each structure contains a milk-secreting lobe (D). D, D, D, Milk-secreting lobes (the surrounding tissues being removed), containing the minute alveoli clustered in bunches, the cells of which secrete the milk. F, F, F, Dilatations of the lactiferous ducts. The alveoli and ducts are lined with epithelial cells.

The secretion of milk may be influenced by the nervous system. It is diminished in mental distress, and increased when the child's mouth touches the nipple or the mother sees or hears her offspring. That the secretion of milk continues after section of the nerves known to supply the mammæ, as claimed by some, must find an explanation in the fact that other nerves furnish fibres to the mammæ not at present known to do so; or, possibly, the vaso-motor nerve-fibres alone exert an influence on the secretion.

The **Nausea and Vomiting** present during one pregnancy and not during another, in the same female, may possibly be accounted for by one or more of four causes: 1. Greater or less activity of the cells of the newly formed embryo. 2. The differences in location of the attachment of the ovum and development of the placenta in the uterus, which may be at the entrance of the Fallopian tube, at the fundus, at any part of the body, or immediately over the internal os, doubtless effect a difference of stimulus on terminals of nerve-fibrillæ. 3. The stretching of the muscular fibres in the uterine wall, affecting the nerve-fibres and their end-organs differently, according to the place of attachment of the placenta. 4. The difference in locality of attachment of the placenta in the uterus may, possibly, cause a difference in the quality and quantity of the lymph absorbed, which, passing into the venous circulation and to the heart and lungs, may, in one or in both of these organs, produce a peculiar stimulation of an afferent impulse through the vagus, and effect reflex (efferent) action through the nervous centres of the origin of the phrenic and splanchnic nerves, influencing the stomach and diaphragm, and so establishing the nausea and vomiting sometimes experienced during gestation.

A **Mole** is a more or less shapeless fleshy mass formed in the uterus. If it be the remains of a degenerated embryo it is termed a *true mole*, otherwise it is termed a *false mole*.

Vesicular Mole—*Vide* Hydatidform Moles.

Hydatidform Moles are numerous watery cysts, or conglomerated vesicles, originating from the degenerated villi of the chorion, each saccule growing from another with a polypoid pedicle from original, or principal, stems (not grape-like). They generally grow alternately—larger from small, and smaller from large—and appear like strings with different sized nodules. Some may loosen and be discharged, while others remain. According to Gooch, "the discharged resemble white currants floating in red-currant juice." Normally, the villi of the chorion grow by a process of gemmation; i. e., by buds. In the case of a mole, the growth is abnormal, owing, possibly, to some fault of local circulation. The cells of each of the villi of which the bud is composed develop abnormally, and originate vesicles by repetition, and so form the hydatidform mass, or false moles. The hydatidform mole is not a true hydatid encysted vesicle, or one closed sac within another.

An **Hydatid** is a vesicle, softer than the tissues of membranes, more or less transparent. It develops not only in the uterus, but also in other organs, though without adhering to their tissue. The formation of a hydatid in the uterus may simulate pregnancy, and its expulsion is generally attended with more or less hæmorrhage.

When the ovum or some of its germs are malformed their derivative cells perform abnormal functions, and produce a mole, which in time is generally expelled from the uterus. It must be remembered, however, that moles are not always the result of impregnation; consequently, in a medico-legal sense, it is of importance to distinguish between the true and false kinds.

Discharges more or less solid may be mistaken for a result of pregnancy. Even in the virgin, squamous epithelium of the vagina, in the form of flakes or tubular casts, which may be thrown off in greater or less bulk; a solid, hard fibrous structure; or a blood-clot may lose its coloring matter, become dense, and enclosed by a membrane; or coagulated blood and lymph enclosed in a sac, very much resemble the remnant of a fœtus. It is difficult to distinguish them from the true mole, and may even require the aid of the microscope. The most obstinate, painful menstruation (*dysmenorrhœa*) is often due to an abnormal and destructive membrane at the menstrual period, which may then pass off in shreds, or as a single mass the size of the entire uterine cavity. It may also occur independently of menstruation, yet is always attended by hæmorrhage and expulsive force, easily arousing a suspicion of pregnancy and

Fig. 78.

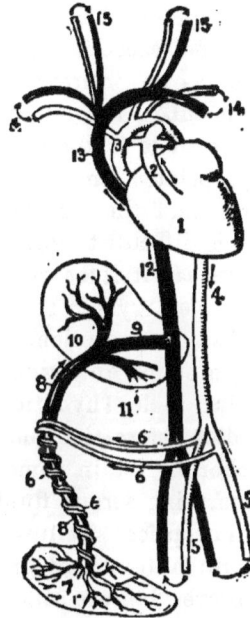

Diagram of the fœtal circulation.

1, Heart. 2, Pulmonary artery arising at the right ventricle. 3, Arch of the aorta. 4, Descending aorta. 5, 5, Internal iliac arteries. 6, 6, Hypogastric (umbilical) arteries. 7, Placenta. 8, Umbilical vein, dividing into (9) the ductus venosus, and (10, 11) branches which enter the liver. 12, Inferior vena cava. 13, Superior vena cava. 14, 14, Vessels to and from the upper extremities. 15, 15, Vessels to and from the brain.

abortion. This is difficult to distinguish from an abortive discharge
if the usual signs of pregnancy are absent and the embryo has
escaped unnoticed, as the dysmenorrhœal membrane resembles the
decidua, with its smooth inner and unequal outer surfaces, that
of non-pregnancy being more ragged, shreddy, and destitute of
cotyledonous sacculi. Again, when the uterine membrane
passes off as an entirety it has the three openings—two of the
Fallopian tubes, and one of the os uteri.

The true mole is the result of conception, the blood supply being appropriated in excess by the decidua and surrounding
structures, and the embryo left to decay from insufficient nutrition. It is difficult to identify a true mole, unless the liquor amnion, or its sac if it be ruptured, or the remains of the umbilical
cord, or the embryonic structures, are found—any of which establish the fact of conception. Moles may co-exist with true pregnancy, the symptoms and expulsion resembling those of abortion. In
a vesicular mole (hydatidform degeneration of the chorion) the
embryo dies before the placenta is fully developed, while the villi
of the chorion retain some of their vitality and continue their
growth, forming serous fluid within. Such growths generally result from dropsical swelling of the villi of the chorion, and resemble in form a bunch of grapes. The size of the vesicular bodies
vary, however, but are attached one to another, being thus distinguishable from true hydatids, which are closed sacs one contained
within another.

In twin conception one ovum may develop normally, while the
other becomes a vesicular or fleshy mole. A mole may remain in
the uterus for a long time—for years even—and it would be haz-
ardous for a physician to assert positively that an embryo, fœtus,
true or false mole, or a true or false hydatid, had been delivered
unless he was present at the time and thoroughly examined the
substance delivered. Moles are generally expelled in a "delivery"
manner, differing in this respect from a tumor or polypus. The
latter may also resemble a mole, but can be distinguished by its
pedicle and the absence of an ovular membranous structure.

Menstruation.—The catamenial, or menstrual, discharges are
commonly known as "courses," "monthly illness," or "periods."
In the female child the Graafian vesicles are small. (See Ovaries.)
After puberty one vesicle enlarges after another in regular sequence,
so that, normally, every twenty-eight days the largest bursts, the
contents passing through the Fallopian tube into the uterus. This

outflow, with the blood from the torn blood-vessels and membrane of the uterus, discharging through the vagina, is known as the *menstrual discharge.* The ripening and dehiscence of Graafian vesicles take place during the child-bearing period. Astonishing freaks of nature are sometimes observed. According to Talten "a woman of the age of 70, a widow for 20 years, was found to be pregnant." Menstruation plays only a secondary part in the phenomenon of ovulation. According to Leishman, a woman who had married at twenty-seven menstruated for the first time two months after her eighth labor; another woman had no discharge until after her second marriage, at the age of forty. Garus refers to a child who menstruated at the age of two years, became pregnant at eight, and lived to an old age. A woman may become pregnant without ever having menstruated, or during the nursing period before the menses reappear.

The menstrual discharge is hæmorrhagic, the blood becoming mixed with the mucus and other secretions of the vagina and adjacent parts. Before and after the rupture of one or more Graafian vesicles, the uterus, ovaries, and Fallopian tubes, especially the mucous membrane of the uterus, become congested, the capillaries distended, its glandular follicles are more active, and a mucus discharge takes place before and after the bloody flux. The proliferation of the cells, and fatty degeneration, together with the congestion, compress the capillary veins, and, in tearing asunder, carry along the capillary vessels and part of the membrane. The epithelial covering and superficial mucous membrane are thrown off, and even blood oozes from the utricular follicles and the Fallopian tubes. The capillary vessels hang denuded in detached loops in the uterine cavity, and may be discovered in the discharge. The slight rupture of blood-vessels of the Graafian vesicles on bursting also adds to said discharge.

Leishman states that Dr. Tyler Smith had several opportunities of examining the uteri of women who had died during menstruation. He found that the appearances presented were similar to those observed after abortion. In one case the mucous membrane in the cervix was perfect, but at the os uteri internum (3, Fig. 72) it ceased as abruptly as if dissected off with a knife above that point, and blood was oozing at numerous points from broken vessels in the submucous tissue. This may be of importance in a medico-legal sense. According to the theory of Reichart and others, "before the ovum is discharged during the process of ovulation the uterine mucous membrane becomes more vascular, congested, spongy, and thicker, in which state it is termed *membrana decidua menstrualis;* that it is then in a more favorable condition to receive, retain, and nourish a fertilized ovum coming in contact with it. If the ovum be not fertilized, then the membrane degenerates and a bloody menstrual discharge follows." According to this view, the discharge is a sign of non-pregnancy. Hence, pregnancy is to be calculated from the time between the last menstruation and the time when it normally should recur. Sometimes, however, ovulation and the formation of the membrana decidua menstrualis occur independently of each other, and menstruation may recur without ovulation, and *vice versa.*

In the opinion of the writer, that process may work differently. During ovulation the extremely vascular, congested, and thickened membrana decidua menstrualis degenerates, the capillary vessels are torn asunder, and bleeding follows. Now comes the point, however, that, when a fertilized ovum is lodged anywhere in the uterus, the uterine membrane is relieved, as it were, as a great part of the blood of the uterine artery is attracted to and supplies the extremely active cells and newly forming tissues of the embryo, when generally no further menstruation occurs. Periodical menstruation may continue, for a time, however, if the capillaries of the membrane should become more or less congested and the membrane degenerate. Again, if the spermatozoon travels the entire length of the Fallopian tube, meeting the ovum near the ovary in the tube-fringe (*m*, Fig. 73) leading from the ovary to the Fallopian tube, and the ovum becomes fertilized in the tube-fringe, then it remains on the spot in the tube, and we have a case of extra-uterine pregnancy. When this occurs the embryo is supplied from the ovarian artery, and menstruation from the uterus may continue as usual from the parts supplied by the uterine artery. (*Vide* Extra-uterine Pregnancy.)

The cause of menstruation is generally associated with, but does not depend on, the maturation of one or more Graafian vesicles and the discharge of ova. Why it recurs in the human female at regular periods of twenty-eight days, when one or more Graafian vesicles mature; and similarly in certain of the lower animals; and in some animals a cluster ripens; and again, in others the ovulation occurs in periods of a year—these are facts which no one can at present explain.

At the cessation of the menstrual discharge in the human female, which occurs generally at the age of about forty-five, the uterus atrophies and the ovaries shrink, denoting the termination of the reproductive function. This crisis in the life of woman is termed the menopause.

Entozoa are animal organisms living within other animal bodies, from whose tissues they derive nourishment; as, for example, worms in the intestines, or trichinæ spiralis in the muscles. The trichinæ spiralis are sexless, encysted, wormlike parasites about 10 millimetres in diameter, found especially in the flesh of animals of the porcine type. If such flesh be used as food (insufficiently cooked) and the cysts digested, the parasites become liberated and multiply in the small intestine, the walls of which they permeate, entering the different parts of the muscular tissue and setting up serious diseased conditions.

Infusoria, or water animalcula, are found generally during the summer months, if the water be exposed in a moderately warm atmosphere, though the water be clarified, distilled, or boiled. The infusoria reproduce their species by eggs in special generative organs, which are fecundated by union of the sexes.

Life Functions pass through unceasing changes until completely suspended, the particular kind, or species of organisms remain, while the individual perishes.

Plants.—In certain plants the two sets of reproductive organs are found together, the ovules being fertilized by the pollen of the same structure. After fertilization the ovules are set free and thrown upon their own resources, to draw from the surrounding elements the materials for their growth and development.

CHAPTER XXVII.

MISCELLANEOUS.

The **Connective Tissues** are of great variety. Their function is that of supporting the frame and holding together the various other tissues and organs. They are divided into mucous and retiform (net-like) connective tissues, white and yellow fibrous tissue, cartilage, and bone. The cells of all these tissues have the property of secreting material for the nutrition and repair of their own and of the intercellular substance. The younger the tissue the more numerous are the cells contained in it; while in old age the intercellular substance predominates, the cellular portion being relatively less.

Fibrous Tissues, especially the non-elastic or white tendinous, are, as a rule, easily affected by chemical reagents. Weak acids cause them to swell and become indistinct; consequently, a super-acid condition of the system causes disease.

Cartilage-cells secrete an intercellular substance, which in some parts remains quite soft, while in other parts it becomes hard.

Bone is, in some respects, the most remarkable of the connective tissue group, its cells having the peculiarity of attracting and secreting a great quantity of earthy, or inorganic, matter. This inorganic matter imparts to bony tissue its great strength and hardness, and enables it to support the body. Bone contains also a quantity of fat, cells, nerves, and blood-vessels, and is covered on the outside with a tough vascular coat, termed the *periosteum*. Where bones are in relation at the joints, their articular surfaces are covered with cartilage. (*Vide* Lime Phosphate.)

The **Periosteum** is a fibrous membrane on the external surface of the bones. It imparts strength to the bone and affords a suitable attachment for muscles. It is well supplied with arteries, thereby furnishing nutrition to itself and the bone. It contains also veins, nerves, and lymphatics. In periostitis, pus may collect between the bone and periosteum, causing pressure on the vessels, so that the outer layers of the bone are soon deprived of nutrition, resulting in necrosis.

Gangrene.—Interruption of the return of venous blood is seldom a cause of *gangrene*. This fact is especially noticeable in women during or after pregnancy, when the veins of the lower extremities are sometimes enormously enlarged. About the only instance where a stoppage leads to gangrene is that of a hernial sac, where the veins are compressed sooner than the arteries. Interruption or entire stoppage of the normal exchange of arterial and venous blood terminates the life of a part (gangrene). A tumor pressing on arteries, or continuous convulsive contractions of their muscular coat, even the continuous use of too large doses of ergot, may produce gangrene. Again, the cause of interruption of the blood circulation may have its seat in the capillaries, as by exudations or new formations in the parenchyma of the tissues of the part affected, producing compression on the blood-vessels. Other causes of a mechanical or chemical kind, such as crushing, concussion, desiccation, corrosion, or ferment poisoning, produce, by their violent disturbance of the molecular arrangement of the tissues, a disturbance which is incompatible with the continuance of vital action in the part affected. (*Vide* Mortification.)

When the amount of arterial blood flowing through a certain part of the body falls below the normal quantity, the nutrition of such part suffers; if the current ceases entirely, so does the nutrition. Disturbances of the circulation are most frequently caused by obstruction in the afferent arteries. This may arise from a thrombus, or disease may cause a gradual diminution of the lumen of the vessel, whereby the heart's propulsive force becomes much reduced in the larger arterial trunks. Disease of the heart's muscle or general debility may also retard the blood current.

In **Gangræna Senilis,** or the dry gangrene of old age, which commences at the fingers or toes and gradually extends upwards, caused by defective circulation, both forces—the heart and the blood-vessels generally—act conjointly; that is to say, there is a retrograde change in the heart's muscle, and disease of the vessels.

New Formations—Cancer.—By a new formation is meant an abnormal increase of development, or excessive growth of tissue. Virchow, writing on cell life, says, "Wherever cells exist in the organism they are the offspring of other cells which no longer exist; the new cells are the heirs of the mother cells, yet not always of their peculiarity and vital properties." Here is condensely expressed the substance of many volumes; for in this new formation the original simple cell is multiplied, the nucleus elongated, and then constricted in the centre until the connection is broken. The nucleus, in dividing, brings about a division also of the nucleolus; then there are two cells instead of one; these two produce four; these eight, and so on *ad infinitum*, if unobstructed. Sometimes this occurs rapidly, as from an injury, causing a tumor or cancerous growth in a short space of time. An increase of protoplasmic force from within outward, or a supply of defective pabulum to the cells, may cause a division of the nuclei, with a similar result. The form of the growth depends upon the the space and kind of tissue it occupies. The cells of such growth are termed *giant cells*, because they are larger than other cells and almost without limit in growth. When imbedded in soft, yielding tissue, the growth produced is generally round; but if imbedded in a fibrous structure, then it is furnished with processes at its periphery, continuing into the soft, yielding interspaces of the fibrous structure—the so-called roots of cancer. This abnormality is an example of the means of unlimited multiplication, primarily, from a single giant cell.

Of all tissues of the body the blood contains the largest percentage of water; therefore, in death it decomposes more rapidly than any other tissue. Soon after death the coloring matter leaves the blood-corpuscles, and colors first the serum, and then the walls of the vessels and loose cellular tissue around them. Soft tissues, on disintegrating, become swollen, the degree of tumefaction, or swelling, depending entirely upon the quantity of water contained therein.

During life there is another increase of the tissues called *fatty degeneration*, wherein the nucleoli of the cells are changed. These nucleoli become enlarged with fat globules, gradually filling the entire cells, which then increase in size three or four times. Those that have been previously round, cylindrical, flat, or of spindle form, assume a different shape, and are termed *granular corpuscles*, being globular aggregations of fat-globules held together by intermediate albuminous substance. When a large number of such granular corpuscles are together the fluid or tissue in whose interstices the granular corpuscles are suspended or deposited is of a yellowish tint; as in *colostrum*, which is a viscid fluid. If left standing, this fluid soon separates into a transparent serous fluid and a cream-like mass consisting almost entirely of granular corpuscles produced in the mammary gland by fatty degeneration, the minute particles of the corpuscles not having been completely broken up, causing the colostrum to appear yellowish or oily. In fatty degeneration of the tissues the albuminous substance referred to is surrounded by the alkaline blood, and its alkaline salts, which are abnormally increased, act upon and dissolve the fat-globules. The albuminous substance is then no longer able to hold the fat-globules together, when they disintegrate or break up. The farther this disintegration of fat-globules proceeds the whiter the surrounding tissue appears, the cells, as previously stated, swelling up and rupturing on account of the overproduction of fat. Rindfleisch says, "Even nerves and blood-vessels become subject to fatty degeneration, the muscles of the heart being most liable."

Dissolution of the Human Body begins as soon as the functional interdependence of the different organs ceases, the parts then becoming like inorganic matter. The only force keeping them in form is cohesion. This, however, exerts a limited and temporary power on human organic matter, owing to the extraordinary abundance of water present. At death decay begins slowly, the process of dissolution accelerating with the lapse of time.

Spontaneous Generation is not admissible even in the case of *bacteria*, including their various varieties. Heat exerts in varying degrees a preventive action on their appearance. *Schizomycetes* (minute vegetable organisms—*saprophytes*, which some claim, belong to the *algæ*) have the peculiarity of being devoid of chlorophyll, and are endowed with mobility in the absence of oxygen. Bacteria, in one variety or another, develop in all kinds of decaying animal (ptomaines) and vegetable matter, and when developed may cause live animal tissues, of greatly diminished vitality, to decay. Organisms described as *micrococcus, bacterium,* and *bacillus* are, possibly, varieties of the same species only.

Ptomaines.—Ptomaine is the principal acting substance of dead animal tissue (*carcass*). It is to the dead tissue what the proteid is to the living. Both proteid and ptomaine differ according to the composition of the tissue. Again, both differ according to the condition of the tissue; for instance, the *proteid* differs much in the plethoric condition of the body from the anæmic. The *ptomaine* differs much according to the decomposition of the dead tissue. In the ptomaine organisms easily develop termed *pathogenytes*, which exercise an important influence in many pathological conditions. Some pathogenytes may be transient, others more permanent and severe. These organisms are very apt to form in a wound, especially in a defective mucous membrane, or in a bed-sore for instance, in which the diminished vitality of the patient favors their formation and absorption. These organisms may easily be-

18

come absorbed by the blood-capillaries direct, or through the lymph-vessels enter and poison the blood, as in the case of diphtheria (*diphtheritic cocci*) from decayed mucous membrane of the tonsils, for instance. When pathogenytes are developed, or introduced with food, the effect is apparent, especially in the intestinal canal, illustrated by the introduction of the *tyrotoxicon* of decayed cheese or other food, and may, in legal investigations, be of great importance in clearing up certain mysterious poisoning cases, which might be mistaken for the willful admixture of poisonous drugs.

As long as the pathogenytes affect a wound only they can be destroyed with antiseptics. Several kinds of said organisms may be combined in one ptomaine, can live and multiply, and by the blood be carried from one part of the tissue to another. They may aggregate in the capillaries of an organ, tissue, or gland, and become absorbed again, producing what is termed **Infection.** The blood-poisoning from wounds is often believed to be due to absorption of pus-cells and is termed *pyæmia*, but in reality very often, possibly always, is due to poisoning from micro-organisms of a ptomaine. When the ptomaine organisms develop quickly within the system, or enter it very numerously by **Contagion**, they may produce death in a few hours. Their slow development and smaller number in the system may produce *hectic fever*. The relapsing, continued, remittent, periodic, and intermittent fevers, respectively, may possibly be due to pathogenytes of ptomaines developed somewhere in the mucous membrane—most likely in the alimentary canal. The time may not be distant when the *primary origin* of Koch's bacillus may be traced not to the lungs, but to defective mucous membrane either in the bronchial tubes, nose, or intestine. It is by no means necessary that the primary origin in all cases be in the same locality.

Æther, or **Ether,** is an extremely subtle, elastic substance which is supposed to pervade all space as well as all bodies. It may reasonably be assumed to be a simple form of gas, which on account of its extreme tenuity is undetectable by any process at present known to science. It does not enter into combination with other substances; therefore its quantity, quality, and action always remain constant. Assuming the atomic theory to be the correct one, æther forms the interstitial medium between the atoms and molecules of all gases, fluids, and solids; entering the inter-atomic and molecular spaces on the expansion of these bodies and receding on their contraction.

Still-born.—In determining whether a dead infant has breathed or not, Moreno says "that the ciliated cylindrical epithelium of the respiratory tract extends down to the terminal bronchioli, where, in an infant that never breathed said epithelium is cubical, but as soon as respiration begins, it is transformed into flat epithelium. That the change not being a mechanical, but a physiological occurrence. Hence, by the aid of a microscope it might be determined whether the infant, whose lungs contain air, had drawn it in by inspiration, or if it had been forced in during an attempt at artificial respiration."

Artificial Respiration is recommended by Sylvester as follows: "The patient is laid on his back on a plane, inclined a little from the feet upward; the shoulders are gently raised by a firm cushion placed under them; the tongue is brought forward so as to project partly from the side of the mouth. The operator then grasps the patient's arms, raising them above the head. This action imitates inspiration. The patient's arms are then turned down and finally pressed for a moment against the sides of the chest. A deep expiration is thus imitated. These two reversing movements should be perseveringly continued at the rate of about 15 to 18 times in a minute. In addition the operator may pay attention to the following: 1. That all obstruction to the passage of air to and from the lungs be at once removed; that the mouth and nostrils be cleared from foreign matter or mucus. 2. A dash of hot water (120° F.) or cold water to the face for the purpose of exciting respiration. 3. That the temperature of the body be maintained by friction or warm blankets. 4. In case of drowning, in addition to the foregoing suggestions, first of all place the body with the face downward, the face hanging a little over the edge of the table or board, and the body raised at an angle of about 30 degrees, so that the head is lower than the feet. Open the mouth and draw the tongue forward, and allow the water from the lungs to escape quickly, which may be assisted by pressure on the back, then to resort to the foregoing artificial respiration."

GLOSSARY.

Abdominal reflex. Contraction of the abdominal muscles upon stimulation (tickling) of the skin over the side of the abdomen or sole of the foot.

Abducens. Sixth pair of cranial nerves.

Abduction. Drawing outward from the median line of the body or of the limb.

Aberration. Wandering from; dispersion of the rays of light in passing through a lens; passage of a fluid into a part not appropriate for it.

Absorbents. Capillary veins and capillary lymphatics.

Absorbent system. Veins and lymphatics.

Abortion. Accidental or willfully induced premature birth.

Abscess. A circumscribed collection of pus in tissue.

Absorption. The taking in or sucking up by the tissues; imbibition.

Acarus. Mite; a minute living organism.

Accessorius. The spinal accessory nerve.

Accessory. Accompanying or taking part with another, as the spinal accessory nerve with the pneumogastric nerve.

Accommodation. The power of the crystalline lens to increase or diminish its convexity so as to focus on the retina the rays of light reflected from objects at varying distances within a range of from (about) 4 to (about) 225 feet from the eye. This accommodation of the lens is caused by nervous reflex action for contraction or relaxation of the ciliary muscle, ciliary and suspensory ligaments of the eye.

Acetate. A salt formed by the union of acetic acid with an alkaline, earthy, or metallic base. (See Base and Salt.)

Acetic. Pertaining to acidity, sourness.

Achromatic. Free from chromatic aberration; destitute of color.

Acidity. Sourness.

Acinous gland. Small gland composed of saccules in the form of a bunch of grapes.

Acinus (*pl.* acini). Ultimate lobule of a racemose gland.

Acoustic. Relating to hearing or sound.

Acrid. Irritant.

Acupuncture. Puncturing the skin, as with needles.

Acusticus. Auditory nerve.

Acute. (*Lat.* acutus). Relating to disease of short duration; sharp; pointed.

Adam's Apple. Pomum Adami, or angular projection on the anterior aspect of the thyroid cartilage.

Adducens (adducent). Relating to adduction.

Adduction. Bringing, drawing, or attracting inward toward the median line of the body, organ, or of the limb.

Adenoid. Resembling a gland or its tissue.

Adenoma. Tumor originating in and composed of glandular tissue.

Adhesion. Attraction between *unlike* molecules. (See Cohesion.)

Ad infinitum. To infinity; unlimited capacity, extent, or energy.

Adipose. Relating to fat; fatty.

Adolescence. The period from puberty to legal majority.

Adventitia. Outer coat of blood-vessels.

Adynamic. Loss of vital power or muscular strength.

Aer. Air; gas.

Aeration. Arterialization.

Aeriform. Like air; gaseous.

Æsthesis. Perception; sensibility.

Æther. See Æther in chapter on Miscellaneous.

Ætiology. The science of causes; doctrine of the causation of disease.

Afferent. Carrying to.

Afferent nerve-fibres. Carrying sensory impulses to either nervous centres of the cerebral convolutions to inform the sensorium, or to excite in other nervous centres reflex (efferent) impulses for motor action of an organ, or for both of these actions.

Affinity. Attraction; relationship; tendency toward.

Afflux. Excessively flowing to.

After-image. A visual impression that persists on the retina after seeing an illuminated object.

Ageusia. Want of taste.

Agglomerate. Crowded together; aggregate.

Aggregation. A collection; an assemblage.

Agglutition. Inability to swallow.

Agminate. Grouped together.

Agminate glands. Peyer's patches.

Ague. Intermittent fever.

Air-cells. Minute dilatations of the terminals of air-passages in the lungs.

Ala (*pl.* alæ). A wing; armpit.

Ala cinerea. Wing ash-colored; vagus area.

Alar. Wing-like.

Albescent. Growing white.

Albuginea. A thick layer of white fibrous tissue investing viscera.

Albuginea ovarii. Tunica albuginea of the ovary.

Albugineous tissue. Fibrous tissue of a yellowish-white color.

Albumen. White of egg.

Albumin. Albuminous (nitrogenous) proximate principle, always containing sulphur.

Albuminate. A combination of albumin with a base.

Albuminoid. Proteid.

Albuminose. Albuminous matter, converted during digestion, fitted to diffuse through animal membranes.

Albuminous. Relates to animal and vegetable food containing albumen mixed with other nutritious substances.

Albuminuria. Albumin in urine; Bright's disease.

Alga (*pl.* algæ). Seaweed; an order of cryptogamic plants.

Algesia. Morbid sensibility to pain.

Aliment. Nourishment; food.

Alimentary. Relating to aliment, or to the alimentary canal.

Alimentary canal. Digestive canal, extending through the body from the mouth to the anus.

Alkali. That which is generally known as soda, potash, or ammonia, and forming a neutral substance with an acid. *Lussac* classifies the alkalies as follows:

I. Metallic bases, as potash, soda, and lithia, containing oxygen.

II. Non-metallic bases, as ammonia, not containing oxygen.

III. Vegetable alkalies or alkaloids, as quinine, morphine, strychnine, etc.; all contain nitrogen in combination with oxygen, hydrogen, and carbon. These are the active principles of certain plants.

Alkaline. Relating to or having the property of an alkali.

Alkaloid. A salifiable base of certain vegetables, existing therein as a proximate principle. (See alkali.)

Allantoid. Condition like force-meat.

Allantois. Chief component part of the placenta, being the internal fibrous layer of the chorion surrounding the foetus and amnion.

Aloe. An inspissated juice of a certain plant.

Alopecia. Loss of hair.

Alveolus (*pl.* alveoli). A small cavity or sac.

Alvine. Pertaining to the abdomen or intestines.

Alvus. Belly.

Amaurosis. Loss of vision.

Ambiopia. Double vision.

Amblyopia. Dimness of vision; partial amaurosis.

Amenorrhœa. Absence of the menses due to some morbid condition.

Amentia. Imbecility of mind; total absence of intellect.

Ametropia. Defective refraction in the eye.

Amide. A compound of an acid radical (atom) with the compound formed by one of nitrogen and two of hydrogen ($N\ H_2$), derived from ammonia. Ammonia thereby loses one radical of hydrogen.

Ammonia. A gaseous compound of one element of nitrogen and three of hydrogen ($N\ H_3$), existing in air and soil, the product of decomposition and putrefaction of organic tissues.

Ammoniacal. Containing or pertaining to ammonia.

Amnesia. Forgetfulness; total or partial loss of memory.

Amnion. A membranous sac forming part of the placenta enveloping the foetus in the uterus.

Amœba (*pl.* amœbæ). A unicellular organism composed of a nucleated mass of protoplasm.

Amœboid. Relating to or resembling amœba, especially in movements and changes of shape.

Amorphous. Without form; shapeless; matter in a solid state when exhibiting non-methodical arrangement of its molecules; not crystalline.

Amplitude. Extent, fullness, strength, especially as applied to sound or pulse.

Amplus. Ample; large; roomy; spacious; great.

Ampulla (*pl.* ampullæ). Oblong vessel or flask, enlarged or puffed out; flask-shaped enlargement of a membranous tube, as that of a semicircular canal.

Amylaceous. Containing or pertaining to starch; starch-like.

Amyloid. A compound formed in the body from starchy matter.

Amylum. Starch.

Anabolic. Relates to anabolism.

Anabolism. Constructive process; building up tissues from nutritive substances; assimilation. (See Metabolism and Katabolism.)

Anæmia. Insufficient number of, or defective, red blood-corpuscles in the blood, or a local or general want of blood in the body.

Anæmic. Relating to anæmia.

Anæsthesia. Impairment or absence of sensation of a part of, or of the entire, body—caused either by disease or by the use of an anæsthetic.

Anæsthetic. A substance or agent that impairs or destroys (paralyzes) sensation; pertaining to or inducing anæsthesia.

Analgesia. Loss of sensation to pain, but not to tactile impressions.

Analogus (analogous). Resembling; similar to.

Analysis. The act of dissolving any substance; separating constituent elements of a compound.

Anasarca. General dropsy.

Anastomose (anastomosis). Union or communication with each other, as of the arteries, veins, lymphatics, or fibres of a nerve with another nerve; inosculation.

Anatomic element. The smallest natural division of the organism.

Anatomy. The art of dissecting organized bodies.

Anchylosis (ankylosis). Immobility of a joint due to abnormal adhesions.

Aneurism. Bulging or tumor-like dilatation of the coats of a vessel; several aneurisms may thus anastomose.

Angina. Severe choking affection.

Angina pectoris. Paroxysms of intense pain in the præcordial region and heart, with a sense of suffocation, diminished pulse, and altered rhythm.

Angioma. Tumor, composed principally of vessels.

Ani. See Anus.

Animalculum (*pl.* animalcul.a). A minute organism.

Annular. Ring-like.

Ano. See Anus.

Antacid. That which counteracts acidity.

Antagonism. Opposition; resistance; counteraction.

Anteflexion. Bending forward, usually referring to the uterus.

Anterior. That which is in front; forward aspect.

Antero. Anterior—used in the formation of compound adjectives.

Anteversion. A displacement forward of the fundus of the uterus.

Antidote. A remedy which counteracts or removes the effects of poison.

Antiphlogistic. A remedy opposed to or which checks inflammation.

Antipyric. A remedy which checks or prevents the formation of pus.

Antiseptic. An agent opposed to putrefaction, or which prevents the growth of organisms in the tissues.

Antispasmodic. A remedy which relieves and prevents spasms.

Antizymotics. An agent opposed to fermentation.

Anuresis. Inability to urinate.

Anuria. Suppression of urine.

Anus (*gen.* ani). The termination of the rectum.

Aorta. The largest artery of the body, originating at the heart.

Aortic. Relating to the aorta.

Aperture. An opening; a passage.

Apex. The summit or extreme point.

Aphasia. Loss or disturbance of the power of speech.

Aphonia. Loss of voice.

Apnœa. Want of breath; extreme inability of respiration.

Apomorphia. An alkaloid prepared from morphine and hydrochloric acid.

Aponeurosis. A thick, white, shining expansion of tendons; a sheet of tendinous fibres.

Apoplexy. Sudden paralysis of sensation and motion, resulting usually from hemorrhage in the brain.

Apparatus. A combination of organs performing harmonious and definite functions, as the circulatory apparatus formed by the heart, arteries, capillaries and veins.

Appendix. An accessory part or a continuation of an organ.

Apposition. The act of bringing together; of being properly together.

Aqua. Water.

Areola (*pl.* areolæ). Interstitial space between fibres of the most delicate connective tissue.

Areola mammæ. The circular brownish color around the nipple.

Areolar tissue. Most delicate connective tissue, having areolæ freely communicating with each other, especially when blown up with air or when distended with fluid.

Argentum. Silver.

Argenti Nitras. A salt prepared from silver, nitric acid, and water.

Argyria. Grayish discoloration of the skin resulting from long-continued internal use of argenti nitras.

Arrector. Lifting up; raising.

Arrectores pili. Erector muscles of the hair.

Arteria. Artery.

Arterialization. The imbibition or taking up of oxygen by the venous blood, making it arterial blood.

Articulation. The approximation of the cartilaginous ends of bones; a joint; also, relating to the proper enunciation of words.

Arytænoid. Funnel- or pitcher-shaped.

Arytænoideus. Relating to arytænoid muscles, cartilage, nerves, etc.

Ascites. Abdominal dropsy.

Asphyxia. Inhibition of respiratory movements; suffocation.

Assimilation. The conversion of nutritious matter into the proper organic tissues of the body.

Asthenia. Debility; want of strength.

Asthenopia. Muscular and accommodative weakness of the eye.

Ataxia. Inco-ordination of muscular movements.

Atlas. First cervical vertebra.

Atom. The smallest part of matter that can take part in a chemical change.

Atony. Diminished muscular tone; want of strength.

Atrophy. Wasting of a tissue, or of the whole or a part of the body.

Atropia. The alkaloid of belladonna.

Attolens. Lifting upward.

Attrahens. Drawing forward or towards.

Auditorius. Auditory nerve.

Auricle. External ear; right auricle of the heart.

Auricles of the heart. The two superior cavities of the heart.

Automatic. Producing movements without external aid; self-acting; self-regulating.

Axilla (*pl.* axillæ). Armpit.

Axis. A right line passing through the center of a body on which it revolves; the second cervical vertebra is so named, from the fact that it forms the pivot upon which the head rotates.

Axis-cylinder. Central filament of a nerve-fibre.

Axis-cylinder process. The prolongation of a nervous centre-cell—believed not to be connected with the axis-cylinder of a nerve-fibre.

Arachnoid. A delicate serous membrane, resembling a spider's web, between the dura mater and the pia mater.

B

Bacillus (*pl.* bacilli); (From *baculum,* a stick or rod). A minute rod; rod-like bacterium.

Bacterium (*pl.* bacteria). A little rod; a micro-organism of rod-like form; genus of schizomycetes, occurring in decomposed animal and vegetable matter.

Basal ganglia. The optic thalami and corpora striata together.

Base. Foundation; resting on; chemically, principal element of a compound; any substance that will neutralize an acid.

Basement membrane. A delicate, structureless lamella beneath the epithelium.

Basilar membrane. The membranous spiral lamina of the cochlea that stretches from the labium tympanicum to the crista basilaris.

Bifurcation. Dividing into two branches; place of division.

Bioplasm. Living matter; protoplasm.

Bipolar. Having two poles or prolongations.

Blastoderm. Layers of cells in the segmented ovum covering the yelk; cells which form the germinative area; germinal membrane.

Bowman's capsules. Envelopes of the glomerules in the kidney.

Bronchial tube (*pl. bronchial tubes*). Tubal continuations from the bronchi to the bronchioli.

Bronchiolus (*pl.* bronchioli). Capillary continuation of the bronchial tube. A single membrane destitute of cartilaginous rings.

Bronchitis. Inflammation of the bronchi.

Bronchus (*pl.* bronchi). One of the two divisions of the trachea.

Bubo. Inflammatory enlargement of a lymphatic gland, especially of the groin or axilla.

C

C. Abbreviation of centigrade and of carbon.

Cachexia. Defective nutrition, as occurring in syphilis and in cancer; depraved condition of the body.

Cadaver. A carcass, or dead animal body.

Cæcum. Ending blindly, as a sac with only one opening.

Calamus scriptorius. A writing pen; a small pen-shaped tubercle at the inferior angle of the fourth ventricle.

Calcareous. Pertaining to lime.

Calcis. Lime.

Calculus. A stone-like formation in the body.

Callosus. Hard; thick-skinned.

Callous. Indurated; hard.

Callus. The flexible bony substance formed around the ends of a fractured bone during the process of repair.

Caloric. Warmth; heat.

Calyx (*pl.* calyces). A cup; cup-like cavity into which the renal papillœ project.

Canalis reuniens. A short canal connecting the ductus cochlearis (scala media) with the saccule.

Canula. A small tube.

Capillary. A hair-like tube or vessel.

Carbohydrate. A compound of carbon, oxygen, and hydrogen.

Carbon. A non-metallic element found in all organic compounds; coal.

Carbonate. A salt formed by the union of carbonic acid with a salifiable base. When the base is imperfectly neutralized by carbonic acid the salt is termed a *subcarbonate;* when there is an excess of acid it is termed a *bicarbonate.*

Carbonic acid. A compound of one atom of carbon and two of oxygen. In the animal body it holds in combination ammonia and organic matter. It has a slight odor and a pungent, acidulous taste, and acts as a powerful stimulus in the animal economy. Carbonic acid gas artificially introduced acts as a sedative.

Carcinoma. A tumor consisting principally of epithelial elements; cancer.

Cardiac. Relating to the heart.

Carotid. An artery on each side of the neck supplying blood to the head; so named because of its slow pulsation during sopor, coma, lethargy, and carus.

Catalepsy. Complete suspension of irritability and volition, with rigidity of the voluntary muscles, the limbs and trunk preserving the different positions given them.

Cataract. Opacity of the eye-lens or of its capsule.

Caudal (Lat. *cauda,* tail). Pertaining to the extremity, or tail-end.

Cellular. Pertaining to or consisting of cells.

Cellulose. Cell-matter and fibres of plants isomeric with starch.

Centigramme. A French weight; the 1-100th part of a gramme, designated 0.01.

Centigrade (*abbr.* C.). A scale to measure temperature, based on the decimal system.

Centimetre. One-hundredth of a metre; 0.3937¹ inches.

Centre. An aggregation of gray nervous cell-matter.

Cerebellum. The posterior lower part of the brain; small brain.

Cerebro-spinal. Pertaining to the brain and spinal cord.

Cerebrum. The upper and largest part of the brain.

Cervical. Pertaining to the neck between the head and chest or to the neck of the uterus.

Cervix. Neck.

Chemical change. A change that alters the identity of the molecule. (See Physical change.)

Chlorophyll. A dark-green and essential element of vegetation, composed of hydrogen, nitrogen, oxygen, and carbon, imparting the green color to foliage and transforming plant-food into vegetable matter.

Chlorosis. Green-sickness characterized by anæmia, languor, palpitation of the heart, and deficiency of red blood-corpuscles, generally affecting young females.

Cholesterin. A fatty substance occurring in the brain, nervous system, and blood; excreted in bile, and occasionally in the urine of fatty kidney.

Chondrin. The essential constituent of cartilage.

Chondrus. Cartilage.

Chorda tympani nerve. A branch of the facial at the Fallopian aqueduct, passing under the mucous membrane of the tympanum (middle ear), then joining the lingual nerve to confer the sense of taste to the anterior two-thirds of the tongue.

Chorea. St. Vitus' dance; irregular involuntary twitching and contraction of muscles.

Chromatic. Pertaining to color.

Chronic. Long-continued, slow, tedious progress.

Cicatrix. A scar upon the skin after the healing of a wound or ulcer. The cicatrix of bone is termed callus.

Cilium (*pl.* cilia). Eyelash; applied, also, to a fine, hair-like process on certain cells.

Cinerea. Ash-colored.

Circumflex. Bending around.

Cirrhosis. Atrophy of normal tissue caused by new formation of connective tissue contracting and obliterating the vessels.

Claustrum. Shut off; an enclosure; layer of gray matter between the island of Reil and the lenticular nucleus.

Cleido. From relation with the clavicle.

Clinoid. Bed-like.

Coccus (*pl.* cocci). A berry; synonym for micrococcus or minute round organism.

Coccygeal. Relating to the coccyx; lowest part of the spine.

Cœliac. Relating to the abdominal cavity.

Cœliac plexus. Solar plexus.

Cogitative. To think; in mind; thought about; power of thinking.

Cohesion. Molecular attraction between *like* molecules.

Coition. A coming together.

Collagen. The essential organic basis of connective tissue.

Colloid. Like gum. Gelatinous and gummy (amorphous) substances do not pass readily through septa, hence the name. (See Crystalloid.)

Coma. Profound sleep with complete unconsciousness and slow, stertorous respiration.

Comedo. Black-heads; face-worms.

Comminuted. Broken into many pieces.

Commissure. A juncture, or point of union.

Concentric. Having a common center; thus, the bulb of the onion is a concentric bulb, consisting of many layers one within the other.

Concept. The mind's own work, or its own power.

Condyle. An articular eminence, round in one direction and flat in the other; a curved extension.

Condyloid. Shape of a condyle.

Condyloma. Soft, fleshy excrescence; a knot; an eminence.

Conglomerate. Crowded together, as of many small glands forming a large gland.

Conjunctiva. Joining together; mucous membrane covering the anterior portion of the eye-ball and the inner surface of the lids.

Conniventes. Keeping; converging towards or approaching one another.

Contaminated. Polluted; corrupt; tainted.

Convolution (gyrus). Winding or folding, especially the outer part of the brain or intestine.

Copula. Shackled, or bound.

Corneous. Horny.

Cornu. A horn.

Coronary. Resembling a wreath or crown; encircling.

Coronoid. Processes of bone resembling a crow's beak.

Corpora pyramidalia. Pyramids of the medulla oblongata.

Corpus (*pl.* corpora). Body.

Corpuscle. A very small body or mass of substance.

Corrugated. Wrinkled.

Cortex. Bark; outer layer of an organ.

Costal. Relating to the ribs.

Cotyledon. A circumscribed tuft of chorionic villi on the attached surface of the placenta.

Cotyloid. Cup-form.

Crepitant. Crackling.

Crescendo. Increase.

Cribrosa (Lat. *cribrum*, sieve). Having holes like a sieve.

Cribriform. Sieve-form.

Cricroid. Ring-form.

Crista acustica. A crest or ridge in the ampullæ of the semicircular canals of the ear.

Crista basilaris. Slight elevation on the ligamentum spirale to which the lamina basilaris is attached.

Cruciform. Like a cross.

Crural. Pertaining to crus.

Crus (*pl.* crura). A leg; resembling a leg.

Crusta. A crust.

Crystalline. Resembling or in part made up of crystals; matter in a solid state exhibiting a methodical arrangement of its molecules.

Crystalline lens. A transparent structure just behind the iris in the eyeball.

Crystalloid. Resembling a crystallus; like crystals; matter that can be crystallized, as salts, acids, and alkalies; crystallizable substances easy to osmose, since they pass readily through septa, hence the name crystalloid. (See Colloid.)

Cupola. Vaulted roof of the apex of the cul-de-sac of the cochlea.

Cupula ampullæ, or C. terminalis. A soft, membrane-like material into which project the nerve-hairs of the crista acustica of the internal ear.

Curare. Same as woorara.

Cutaneous. Pertaining to the skin.

Cyst. A circumscribed membranous cavity occurring abnormally in the body, containing fluid, fat, or solid matter, but rarely pus.

Cystic. Containing or composed of cysts; pertaining to a cyst, urinary, or gall bladder.

Cysticercus. The sexually immature larval form of tænia, or tapeworm; tailed bladder worm; hydatid.

D

Deciduus, a, um. Falling off; membrane surrounding the embryo.

Decussation. Crossing.

Defecation. Evacuation of the bowels.

Deferens. Bearing away; transporting.

Deglutition. Swallowing.

Dehiscence. A bursting or splitting open.

Deliquesce. To become pasty and melt by absorption of water.

Delirium. Irregular mental action, with illusion or hallucination.

Dementia. Loss of reasoning power; incoherency of ideas; last stage of insanity.

Dementia congenita. Idiocy.

Dentatus, a, um. Toothed.

Denticulate. Having small teeth.

Denuded. Laid bare; dissected off.

Desquamation. Peeling or scaling off.

Deterioration. Degenerating; impairing; growing worse.

Diabetes. To pass through.

Diabetes mellitus. A disease with sugar in the urine.

Diagnose. The act of recognizing a disease by its symptoms; or, to distinguish one disease from another.

Dialysis. The process of separating mixed liquids or fluid substances by osmose; *i. e.*, passage through an animal membrane.

Diaphoretic. A remedy increasing secretion of sweat.

Dirigent. Director; regulator; harmonious movements.

Diuresis. Excessive discharge of urine.

Diverticulum (*pl.* diverticula). A tube branching out of a longer one; a hollow appendage attached to and communicating with the intestinal tube.

Dorsum (*adj.* dorsal). The back.

Duct. A passage or canal.

Durus, a, um. Hard. **Dura mater.** The external, dense, fibrous membrane surrounding the brain and spinal cord.

Dynamia. Vital force.

Dyscrasia. A bad constitution; morbid condition of the system, of the blood, or of a secondary nature as from a disease.

Dysphasia. Difficult speech.

E

Earthy phosphates. Phosphate of ammonia, magnesia, and lime, found in urine.

Economy. The aggregate of the laws which govern the organism; the entirety of an organic system.

Efferent. A vessel transporting fluid, or a nerve conveying an impulse, away from an organ.

Element. An original, essential part or principle of which a gas, fluid, or solid consists; *i. e.*, their smallest natural division, which cannot be further divided without destroying the identity, such as hydrogen, sulphur, carbon, oxygen.

Emetine. The active principle of ipecacuanha.

Eminentia. A projection, process, or prominence.

Emission. Discharge or throwing out.

Emphysema. Air-swelling or tumor-like infiltration of air into areolar tissue.

Empyema. Collection of pus in the pleural cavity. Sometimes applied to a similar condition in other parts.

Emulsion. A preparation of an oleaginous or resinous substance with the addition of gum or the yelk of egg, held in suspension in a watery fluid.

Encephalon. The brain and nervous matter within the cranial cavity.

Endocardium. The inner lining of the heart-cavities, continuous with the endothelium of the inner coat of the great blood-vessels.

Endogenous. Growing by multiplication or enlargement of the inner cells or tissues.

Endomysium. The extension of the perimysium to and between the muscular fibres.

Endoneurium. The minute web of connective tissue between the nerve-fibres within the nerve.

Enema. A rectal injection.

Ensiform. Sword-shaped.

Enteric. Pertaining to the intestine.

Ephemera. A fever of short duration.

Epigastric. Over the stomach; relating to the *epigastrium;* abdominal wall in front of the stomach.

Epiglottis. A thin fibro-cartilage covered with mucous membrane at the upper part of the larynx, behind the base of the tongue and serving as a cover during deglutition to prevent food from entering the trachea.

Epilepsy. A disease occurring by paroxysms, with convulsive movements of the voluntary muscles and loss of consciousness.

Epithelioma. Carcinoma originating in epithelium.

Erectile tissue. A tissue capable of turgescence and increase of size.

Erythema (*pl.* erythemata). A hyperæmic, red, circumscribed, non-elevated patch on the skin.

Escharotic. A substance by which tissue may be destroyed, leaving a slough or crust.

Etherization. Ether as an anæsthetic; influence of ether.

Ethmoid. Perforated, as a sieve.

Etiology. Doctrine of causes of disease.

Euonymin. The principal active substance of euonymus.

Evacuation. A discharge of matter; an emptying.

Exanthemata. An infectious, febrile, eruptive disease.

Excoriation. An abrasion, especially of the skin.

Excrement. Matter naturally discharged from the body as useless.

Excretion. Thrown off or cast out.

Extravasation. Escape of fluid from its natural place to approximate tissues.

F

F. Fahrenheit scale to measure temperature.

Facet. A separate articular surface.

Fæces. Alimentary excrementitious matter.

Faradization. Application or effect of the Faradic electric current.

Fascia. A bandaged bundle; especially the connective sheath-appendage over muscles.

Fasciculus (*pl.* fasciculi). Bundle of fibres.

Fauces. The short passage between the back part of the mouth and the larynx and pharynx.

Feces. See Fæces.

Fertilization. Rendering or imparting the power to grow and develop; fruitful.

Fibre. Thin; thread-like.

Fibrillæ. Minute fibres.

Fibrinogen. The principal constituent of fibrin.

Fibro. Prefix to words referring to fibrous tissue, either by composition, derivation, or resemblance.

Filament. A minute thread or fibre.

Fissure. A depression or fold inward; a cleft; narrow chasm between partially separated tissue.

Fistula. An abnormal narrow canal with one or two end openings and extending from one organ to another; incomplete or blind with only one opening, but complete with two.

Flexure. A bend or curve.

Fœticide. Killing of the fœtus; criminal abortion.

Fœtus. Product of conception after three months—previously termed embryo.

Follicle (Lat. *follis*, a small bag). A small tubular secreting sac.

Folliculus (*pl.* folliculi). Follicle.

Foramen (*pl.* foramina). An opening; a hole.

Fornicatus. Arched or vaulted over.

Fossa. A depression; a shallow, wide-mouthed cavity.

Fovea. A small, shallow depression or fossa.

Function. The action of an organ or set of organs.

Fungus (*pl.* fungi). Mushroom; an order of cryptogamic plants.

Funiculi. Bundles of nerve-fibres; small, cord-like structures.

Fusiform. Spindle-shaped.

G

Ganglion (*pl.* ganglia). Small lumps or masses of special organs in the body, appearing under three forms: Those of the nervous system are always connected with nerve-fibres; the lymphatic (glands) with the lymphatic vessels; other ganglia, such as the thymus, thyroid (glands), and the suprarenal capsules.

Gasserian ganglion (so named in honor of Dr. J. L. Gasser). This ganglion is on the trifacial nerve, and rests upon the apex of the petrous portion of the temporal bone, and is enclosed between two sheets of the dura mater.

Gaster. Stomach; belonging to the belly or abdomen.

Gastric. Relating to the stomach.

Gastro. Prefix relating to the stomach.

Gastro-intestinal. Relating to the stomach and intestines.

Gemma. Bud; granule; granulation.

Gemmation. Production by budding. On the parent organism buds grow, then separate into complete organisms.

Genesis. Generation; production.

Genio. Relating to the chin.

Genital. Pertaining to generation or to the generative organs.

Genito-urinary. Relating to the generative and urinary parts.

Germ. A minute organic mass, capable of developing into a cell, organ, or organism.

Germ-cell. The cell which contains the germ for reproduction of an organism; ovum-cell.

Germination. Development of germs; vital development; first growth.

Gestation. Bearing or carrying; pregnancy.

Giant-cell. An abnormally large and active cell, originating from an abnormally increased number and size of nuclear divisions of protoplasm of the parent cell.

Gland. An organ composed of cells, blood-vessels, nerves, and absorbents, with the function either to secrete fluid for use in the economy or to excrete as waste.

Glandula (*pl.* glandulæ). A small gland.

Glenoid. A shallow cavity; socket- or pit-like.

Globule. A small ball.

Globulin. An albuminoid of blood-corpuscles.

Globus. A ball.

Glomerule. A small ball-like mass.

Glossa. Tongue.

Glosso-pharyngeal. Relating to the tongue and pharynx.

Glottis. Upper opening of the larynx at the posterior end of the tongue.

Glucose. Grape-sugar; starch-sugar.

Glucoside. An organic compound capable of the production of glucose.

Gluten (*gelo*, congeal). The viscid, elastic, nitrogenous constituent of flour.

Glutin. Vegetable albumen or gelatin.

Glycocholic acid. A principal acid of bile.

Glycosuria. Sugar in urine.

Goblet cell. Goblet-shaped cell found among epithelial cells lining the mucous membrane, with a broad, open extremity.

Goitre. Chronic enlargement of the thyroid gland.

Gonorrhœa. A specific, contagious inflammation of the urethra or vagina, with a muco-purulent discharge.

Graafian vesicles or follicles. Small sacs enclosing ova in the ovary.

Gracilis. Slender.

Granule. A small grain-like particle or body.

Gravidus. Pregnant; impregnated.

Graviditas. Pregnancy.

Gustation. Tasting.

Gynæcology. Science of diseases peculiar to women.

Gyrus (*pl.* gyri). Convolution.

Gyrus fornicatus. Convolution of the corpus callosum.

H

H. Symbol for hydrogen.

Hæmatemesis. Vomiting blood.

Hæmaturia. Blood in the urine.

Hæmoglobin. A crystallizable hæmatin and globulin substance, originating in and coloring the red blood-corpuscles.

Hæmoptysis. Coughing blood from the lungs or air-passages.

Hectic. Consumptive; pertaining to a wasting disease; fever of phthisis.

Helicotrema. Snail-hole; orifice of communication at the summit of the cupola at the junction of the vestibular and tympanic passages of the cochlea.

Helix. Bent; turned round; curled inward, as the external border of the outer ear.

Helosis. Turned outward.

Hemi. Half.

Hemianæsthesia. Impairment or loss of sensation of one lateral half of the body.

Hemiplegia. Paralysis of one lateral half of the body.

Hemisphere. Half a globe; one lateral half of the brain.

Hepar. Liver.

Hepatic. Belonging to, pertaining to, or resembling the liver.

Hepatization. Conversion into a substance resembling the liver; a state of the lungs when engorged with effused matter (as in pneumonia or pneumonitis); no longer pervious to inhaled air.

Hernia. Rupture.

Hibernation (Lat. *hibernare*, to winter). The sleeping of certain animals during the entire cold period of the winter.

Hilum, or hilus. The external depression where the nerves, ducts, and vessels enter and depart at certain organs, as on the kidneys.

19

Hippocampus. Resembling in shape the downward, bended head and appended tail of a horse.

Histology. Science of the minute anatomic constituents of organs and tissues.

Homogeneous. Of the same kind, nature, or structure.

Humor. Animal fluid; a secretion.

Hyaline. Transparent; glass-like.

Hyaloid. Relating to hyaline.

Hybrid. Offspring of two animals or plants of a different species, and not capable of propagating its own species.

Hydatid. A tumor or vesicle consisting of a membrane filled with pellucid fluid.

Hydra. A minute fresh-water polyp.

Hydration. Adding water.

Hydrocarbon. A compound of hydrogen and carbon only.

Hydrocephalus. Accumulation of water in the brain; serum within the ventricles or between the meninges of the brain; dropsy of the brain.

Hydrogen. The lightest gas known — about 14½ times lighter than the atmosphere. In combination with oxygen it forms water. Hydrogen is an element of all organic bodies.

Hydrophobia. An acute fatal disease resulting from the morbid poison of the saliva of a rabid animal, generally received from a bite.

Hydrothorax. Accumulation of serous fluid in the pleural cavity.

Hygroscopic. Able to absorb moisture.

Hymen. Thin fold of mucous membrane at the entrance to the vagina in the virgin.

Hyperæmia. Increased quantity of blood in the capillary vessels of a part; congestion.

Hyperæsthesia. Excessive sensibility.

Hyperplasia. Excessive increase of elements of a part.

Hypertrophy. Increase of a part from excessive nutrition, without change of the nature of the constituents.

Hypochondriac region. The anterior sides of the abdominal cavity below the ribs to a transverse line from Poupart's ligament.

Hypodermatic. Under the skin.

Hypogastric. Below the stomach; abdomen.

Hypoglossal. Under the tongue.

Hysteria. A neurosis occurring in paroxysms, with abnormally increased sensations; excessive susceptibility, with loss of voluntary power to restrain.

I

Icterus. Jaundice.

Idiopathic. Primary; not derived from nor depending on another.

Ileo. Relating to the ileum.

Ileo-cæcal valve. A double fold of mucous membrane guarding the opening from the ileum into the large intestine.

Ileum. Lower portion of small intestine, extending from the jejunum to the ileo-cæcal valve.

Ilia. Flanks.

Iliac. Relating to the ilia.

Ilium. Superior expanded portion of hip-bone.

Imbibition. Sucking in or absorbing.

Inanition. Starvation from defective or scanty nourisment, or disorders in the nutritive processes.

Incontinentia. Incontinence; inability to retain a natural evacuation.

Indolent. Painless tumor of slow growth, or ulcer with no tendency to heal.

Indurated. More solid than normal; hardened.

Inertia. Loss of power to contract.

Inflammation. Heat, redness, swelling, and pain in a part.

Influenza. A special epidemic, catarrhal fever, with inflammation of gastric, œsophageal, and bronchial mucous and olfactory membranes, with nervous pains and prostration.

Infundibulum. Funnel-shaped.

Infusoria. Minute ciliated organisms; micro-organisms; protozoa.

Ingesta. Thrown in; contents of the alimentary canal, as food and drink.

Ingravidation. Impregnation; pregnancy.

Inguinal. Relating to the groin, or region where the thigh joins the body.

Inhibition. Momentary diminishing or stopping of organic activity through nervous influence.

Inorganic. Without organs; that which never had life.

Inosculation. Union; communication; anastomosis.

Insomnia. Sleeplessness.

Interarticular. Between joints or articulations.

Intercellular. Between cells.

Intercostal. Between ribs.

Interlobular. Between the lobules of a gland.

Intermittent. Ceasing and returning at regular or uncertain periods, as the ague fever, and others.

Interstitial. Between tissues; pertaining to connective tissue.

Intertrigo. Chafing or excoriation between two integumental fleshy parts.

Intravasation. Entrance of abnormal fluid into a perforated blood- or lymph-vessel.

Intravascular. Within a blood-vessel.

Introversion. Turned inward.

Ischium. The lower part of the hip-bone.

Isomeric. Having the same chemical formula, but possessing different physical and chemical properties.

Isthmus. The narrowed portion of an organ or passage connecting two organs or passages.

J

Jugular. Pertaining to the throat or jugular vein.

Jugulum. Pertaining to the *jugum*, because fastened to the throat or anterior portion of the neck.

Jugum. A yoke.

K

Katabolism. Destructive process; tearing down tissues; degeneration. (See Anabolism and Metabolism.)

Keloid. An irregular, reddish, firm, circumscribed, elevated connective-tissue growth on the skin.

Kinetic. Effecting or involving motion.

Kinetic energy. Energy which is essential to, or possessed by, a body or substance. (See Potential energy.)

Kreatin. A white bitter substance produced during tissue waste of muscles, brain, and blood.

Kreatinin. A normal nitrogenous constituent of urine, derived from kreatin.

L

Labium (*pl.* labia). Lip.

Labyrinth. A structure of intricate winding passages; the internal ear formed by the vestibule, cochlea, and semicircular canals within the petrous portion of the temporal bone.

Lac. Milk.

Lacerated. Torn or mangled.

Lachrymal. Pertaining to tears; the lachrymal artery, cells, duct, fossa, gland, groove, nerve, papilla, sac, tubercle, and veins.

Lacteal. Milky; also, pertaining to the lacteal vessels.

Lacteals. Those lymphatic vessels that collect the chyle in the intestines.

Lactic. Pertaining to or derived from milk.

Lactose. Milk-sugar.

Lacuna. A small opening, pit, or depression, especially of a duct on the surface of the mucous membrane.

Lamella, or Lamina. A thin sheet, layer, or plate.

Lanugo. The soft, fine, rudimentary, woolly hair on the body.

Laryngeal. Pertaining to the larynx.

Larynx. The upper part of the trachea, or windpipe, containing the vocal chords.

Laxator. Having a loosening or relaxing power, such as that produced by certain muscles.

Lead-palsy. Paralysis of the extensor muscles of the forearm.

Lens. See Crystalline Lens.

Lesion. Injury; derangement; morbid change in structure or function of an organ or tissue.

Lethalis. Mortal; deadly; deep stupor.

Lethargy. Heavy, constant sleep without a conscious interval of waking; deep stupor.

Leucin. White; a crystallizable, fatty product, from the decomposition of albuminous (nitrogenous) animal matter, often found in urine from disease of the liver, typhoid fever, small-pox, or muscular degeneration.

Leucocytes. Colorless, small, cell-like, spherical and nucleated corpuscles of protoplasm, having the power of amœboid movements, floating in lymph and blood, derived from the lymph-glands of the mesentery.

Leucocythamplio. Increased number of leucocytes, accompanied by an abnormally decreased number of red blood-corpuscles.

Leucomaines. Poisonous animal alkaloids developed in the living body by metabolic processes; constituents analogous to ptomaines.

Leucorrhœa. White discharge, especially from the female genitals.

Levator. Elevator; a muscle which raises the part to which it is attached.

Life. Vitality; that enabling metabolism.

Ligament. A white, strong, fibrous structure investing the joint, or uniting bones, or retaining an organ in its position.

Ligate. To tie, bind, or constrict a part of an organ.

Ligature. The appliance used in ligation, or binding.

Limbic lobe. The convolutions above and on the side next to the corpus callosum.

Limbus. An edge, or border.

Line. One-twelfth of an inch.

Lingua. Tongue.

Liquefaction. Becoming liquid.

Lobule. A small lobe.

Locus. Place; spot.

Locus niger (substantia nigra). A layer of dark ganglionic cells in the crura cerebri, separating the crusta from the tegmentum.

Logarithm. Referring to equality of force, strength, account, proportion, and number.

Lucid. Bright; clear.

Lucidus. Bright; light; glittering; sparkling.

Lumbar. Pertaining to the loins.

Lumbar region. Sides of the abdomen.

Lunatic. An insane person with lucid intervals.

Lunula (dim. of *luna*, moon). A small crescent.

Luteus, a, um. Yellow.

Lymph. The fluid within the lymphatic vessels.

Lymphatic. Pertaining to lymph, lymph vessels, or glands.

M

M. Metre.

MM. Millimetre.

MMM. Micromillimetre.

Macula. A spot or stain, without depression or elevation.

Magnus, a, um. Great; large; powerful.

Major (from *magnus*). Bigger; greater; more powerful.

Malic acid. Acid of fruit.

Malignant. Dangerous to life; cancerous.

Malpighi. Relates to parts first described by Dr. Malpighi, an Italian anatomist of the sixteenth century.

Mamma (*pl.* mammæ). A breast; mammary gland.

Mammillary. Nipple-like; relating to mamma.

Mania. Insane excitement; hallucination and delirium.

Manubrium. Hilt, haft, or handle of anything.

Mass. Any quantity of matter that is composed of molecules.

Masseter. Chewing; assisting in chewing.

Mastoid. Nipple-like projection.

Mater. Mother.

Matrix. A mother, womb, or place of origin of propagation of an animal, of tissue, or any substance, but never applicable to a woman or the womb of a woman.

Matter. Anything that occupies space or takes up room, or that can exert or be acted on by force. Matter may exist in masses, molecules, or atoms, in either solid, liquid, or gaseous form.

Maxilla. Jaw.

Meatus. Canal.

Medulla. Marrow; marrow-like.

Meibomian glands. Small sebaceous glands on the margin of the eyelids, between the conjunctiva and the cartilage, first described by Dr. Meibomius.

Meta. A prefix signifying beyond, over, after.

Metabolism. Process of change, alteration, metamorphosis. It includes anabolism and katabolism. (See Anabolism and Katabolism.)

Metamorphosis. Change of form, shape, function; transformation.

Metaphysics. The science of things above and beyond physics; generally employed as synonymous with mental philosophy.

Metre. A French measure, being 39.371 inches.

Micrometer. A scale for minute measurement under the microscope.

Micromillimetre. The millionth part (0.000001) of a millimetre.

Micron. The one-thousandth part (0.001) of a millimetre.

Micro-organism. A very minute organism.

Micturition. The discharge of urine from the bladder.

Millimetre. The one-thousandth part (0.001) of a metre.

Mitral. Resembling a mitre—*i. e.*, a bishop's hat; name applied to the left auriculo-ventricular valve.

Modiolus (dim. of *modius*, measure). The central pillar of the cochlea of the internal ear.

Molecular attraction. See Cohesion and Adhesion.

Molecule. A compound of atoms; the smallest compound of matter that can exist by itself; the physical unit of matter.

Mollis. Soft; elastic; delicate.

Monomania. Insanity confined to a single idea or subject.

Morphology. The science of the forms and elementary constituents of tissues, organs, and organisms.

Morsus (Lat. *mordeo*). A bite; grasp; sting.

Mortification. Loss of vitality. When the part of a soft tissue mortified is recoverable, it is called *gangrene;* if totally destroyed, or dead, *sphacelus.* Mortification of bone when recoverable is called *caries;* when totally destroyed, *necrosis.*

Mucin. The sticky, adhesive substance of mucus.

Muco-purulent. Containing mucus and pus.

Mucosin. Mucin.

Mucous (Lat. *mucosus*). Of the nature of mucus; relating to mucous membrane.

Mucus. A slimy secretion of the mucous membrane.

Muscularis mucosa. A thin layer of unstriated muscle-fibres forming part of the mucous membrane.

Musculin. Chief albuminoid constituent of muscle.

Myo. Relating to muscle.

Myolemma. A thin transparent sheet surrounding the muscular-fibre; sarcolemma.

N

N. Stands for nitrogen.

Nabothian glands, or ova. Small, yellowish glands having follicles with no orifice, occasionally distended by their secretion, found between the folds of mucous membrane lining the cervix uteri. An anatomist named Naboth, finding them morbidly enlarged, mistook them for ova, hence their title.

Narcotic. A substance producing lethargy or stupor.

Necrosis. Death of bone. (See Mortification.)

Nephritis. Inflammation of the kidney.

Nerve-fibrillæ. The minutest nerve-fibres within and between nervous centres. Many fibrillæ together appear like marrow.

Nervous. Relating to the nervous system, nervous centres, their stimulation, action, excitation, and disturbances.

Nervus octavus. Relates to certain nerve-fibres contained in the nervus acusticus (auditory nerve), controlling muscular equilibrium of the head and body, but not connected with the sense of hearing.

Neural. Relating to a nerve or nervous system.

Nitrogen. A colorless, tasteless, inodorous, incombustible gas, 78 parts of which, with 22 of oxygen, constitutes atmospheric air. It is the generator of nitre, hence its name.

Nucleolus (*pl.* nucleoli) The small nucleus within the nucleus of a cell.

Nucleus (*pl.* nuclei). A central differentiated part of the cell; a nut, or kernel within a nut.

O

Occiput. The posterior inferior portion of the head.

Oculomotor. Mover of the eye.

Oculus. Eye.

Odontoid. Tooth-like.

Œdema. Swelling of tissue from an accumulation of serous fluid therein.

Œsophagus. That portion of the alimentary canal extending from the pharynx to the stomach.

Olfactory. Pertaining to the sense of smell.

Omentum. A doubled membranous layer extending from the peritoneum to, and floating or laying over, a portion of the intestines.

Ophthalmic. Relating to the eye.

Os. Bone; also mouth, as *os uteri*, mouth of the womb.

Oscillation. Vibration.

Osseous. Bony, or resembling bone.

Ossicle. A small bone.

Ossification. Formation of bone.

Otic. Relating to the ear.

Oxidation. Conversion into an oxide, as of metals or other substances.

Oxide. Neutral or basic compound of oxygen with a metalloid.

Oxygen. A tasteless, colorless, inodorous gas, 22 parts of which, with 78 of nitrogen, constitutes atmospheric air. One part by volume of oxygen with two of hydrogen constitutes water.

P

Pabulum. Aliment; food. The term pabulum may properly be limited to the food of cells; *i, e.,* the nutritious substances entering the organism become gradually more and more changed, and on arriving at the cells of tissues become protoplasm, or life matter.

Palmaris. Referring to the palm of the hand.

Papilla. Small conical eminence; nipple-shaped.

Para. A prefix signifying beside, beyond, in addition.

Parenchyma. The substance of glandular and other organs, composed of agglomerated globules united by cellular tissue, such as the lung, liver, spleen, and kidney.

Parietal. Side wall of a cavity, especially the side of the head.

Parotid. About or near the ear; relating to the salivary gland situated near the ear.

Patellar reflex. Knee-jerk.

Pathetic. Feeling; sympathy; name applied to the fourth pair of cranial nerves.

Pathogenesis. Production of disease.

Pathogenytes. Minute organisms which can grow in the living tissues, lymph, and in the blood; schizomycetes.

Pathology. The doctrine of disease, its nature, and results.

Pedicle. Stem; neck; stalk.

Peduncle. Stalk or neck-like process by which an organ is attached.

Pellucida. Clear; transparent.

Pelvis. Basin; large, irregular-shaped bone forming a conoidal cavity, open above and below, and enclosing the intestines, urinary and genital organs, its upper portion forming the hips.

Peri. Prefix signifying around or about.

Pericardium. A double membranous sac enclosing the heart.

Perineum. The space at the inferior region of the trunk, between the ischiatic tuberosities, anus, and genital organs, smaller in the female than in the male, triangular in shape, and divided into two equal parts, by the *raphe*. It is occasionally ruptured in labor. The part between the pudendum and anus is called anterior perineum, to distinguish it from the part extending from the anus to the coccyx, called posterior perineum. Both the anterior and posterior are marked by the *raphe* (suture of the skin).

Periosteum. Fibro-vascular membrane surrounding or lining the bones.

Peripheral. Pertaining to the outer surface.

Peristaltic. Rhythmical vermicular motion of the intestine propelling its contents, termed peristalsis or peristaltic motion.

Peritoneum. A strong extensive serous membrane surrounding all the viscera of the abdomen, lining the diaphragm, and through its orifices communicating with the pleura above. Below it lines the abdominal cavity; anteriorly it is doubly reflected, forming the omentum, also the mesentery and ligaments of various organs; it passes over and surrounds the greater part of the bladder and uterus.

Peritonitis. Inflammation of the peritoneum.

Perversion. Turning or changing from good to bad; depravation; taking the wrong course.

Petrous. Stone-like; referring to the hard, stone-like portion of the temporal bone in which is situated the organ of hearing.

Peyer. A Swiss anatomist who first described Peyer's glands.

Phagedenic. Gangrenous.

Pharyngeal. Belonging to the pharynx.

Pharynx. Cavity between the posterior part of the tongue, nasal cavity, and œsophagus; the upper portion of the food channel.

Phonation. Voice producing.

Phosphate. A salt of phosphoric acid.

Phrenic. Diaphragmatic; relating to the diaphragm.

Phthisis. Tuberculosis; any pathological process causing continuous change and destruction of the lung or other tissue.

Physical change. A change not altering the identity of the molecule (See chemical change).

Pia Mater (tender mother). The vascular, delicate, plexiform membrane or immediate covering of the brain, penetrating its anfracuosities, and the continued envelopment of the spinal cord.

Pigment. Coloring matter.

Pilocarpine. The active principle of jaborandi; a powerful sudorific.

Pilus (*pl.* pili). A hair.

Pinna. Wing-like; broad part of the external ear.

Pituitary. Slimy; phlegm-like; relating to the secretion of mucus.

Planta. Sole of the foot.

Plastic. Submissive to formation or nutrition; yielding.

Plethora. Excessive fulness of the blood-vessels.

Pleura. The closed serous sac lining the internal surface of the thorax, divided by a septum (the mediastinum), each sac surrounding a lung.

Plexus. A network.

Pneumaticus. Windy; relating to air.

Pneumogastric. Pertaining to the lungs and stomach.

Pneumothorax. Accumulation of air in the cavity of the pleura.

Polypus. A tumor on the mucous or serous membrane within a natural cavity, attached by one or more pedicles, occurring generally in the nose, vagina, uterus, or rectum.

Pons. Bridge.

Pons Varolii. A bridge-like structure of the brain described by Dr. Varolius. It is a thick, arched band lying across the brain-stem between the medulla oblongata and the crura cerebri. The nerve-fibres from the brain downward, and the spinal cord upward, pass through the pons.

Portal (*porta*, door). Relating to the portal vein or portal system.

Portio dura. Hard portion; the facial nerve.

Portio mollis. Soft, tender portion; the auditory nerve.

Potash. Vegetable alkali, obtained in an impure state by the incineration of a vegetable.

Potential. Having latent power, which may or may not be used in future.

Potential energy. Dormant energy, requiring favorable conditions for its manifestation; accidental energy. (See Kinetic energy.)

Præcordia. Parts in front of the heart.

Prima via. The alimentary canal from mouth to anus.

Principle. The chief part; fundamental substance.

Prism. A rectangular body having generally three rectangular plane faces, or sides, and two rectangular ends.

Process. A prolongation or eminence connected with the principal part of an organ.

Procidentia. Prolapse.

Prognosis. The knowledge of the course and termination of a disease.

Proliferation. Reproduction; cell-division; budding; gemmation.

Prolific. Productive; abundant.

Prominentia. Eminence; protuberance.

Prophylactic. Preventive.

Prostate. Standing before.

Protoblast. Protoplasm.

Protoplasm. Matter that is alive; the essential to the phenomena of all organic life; cell matter.

Protozoa. Single-cell animalcula.

Protuberance. Eminence; projection; pons Varolii.

Proximate. Next in order; nearest to.

Psychical. Relating to the soul or mind.

Psychical blindness. Soul-blindness.

Psychology. Doctrine of the human soul or mind.

Ptoma. Corpse; carcass.

Ptomaine. The principal acting substance in decayed or putrefied animal or vegetable tissue or matter. The ptomaine is to dead tissue or matter what the albuminoid (proteid) is to the living.

Pubic. Relating to the pubes.

Pubes. The bony arch above the external parts of the genital organs.

Pudenda. External genital organs of the female.

Pudic. Relating to the external genital organs and the parts immediately above.

Puerperal. Relating to childbirth or a consequence thereof.

Pulmonary. Relating to the lungs.

Purulent. Containing or producing pus.

Putrefaction. Decomposition of dead organic matter.

Pyæmia. Fever due to pus absorbed into the blood.

Q

Quadrigeminus (from *quatuor*, four). Fourfold. The corpora quadrigemina are so called because they are like one corpus in nature and functions, but constitute four protuberances—two in each hemisphere.

Quasi. As if; having the resemblance of something.

R

Racemose. Clustered; in bunches.

Rachis. Vertebral column; spine.

Rachitis. Inflammation of the spine. Defective ossification with consequent distortion and bending of the bones; the rickets.

Radiation. The transmission of energy through the medium of ether.

Rale. Rattle; crepitation; rhonchus.

Ramification. Branching.

Rarefaction. Expansion.

Receptaculum. A receptacle.

Refraction. Turning or deviating from a direct course on entering a medium of different density.

Region (physiologically). The tract; extent; or area.

Regurgitation. Flowing back; reflux.

Renal. Relating to the kidneys.

Resonance. Increase of sound by echo or vibrations of the walls of a cavity.

Restiform. Cord or rope-shaped.

Rete. Net; interlaced with fibres, vessels, or nerves.

Retiform. Net-like.

Retrahens. Drawing backward; name of a muscle on the back part of the external ear.

Retroflexion. Bending backward, especially of the uterus.

Retroversion. Turned backwards; term applied to that position of the uterus where the fundus is inclined toward the sacrum and the os toward the pubis.

Rhachitis. See Rachitis.

Rhomboid. Four-sided, with opposite angles equal, but with the front more exposed.

Rhonchus. A rattling or wheezing sound.

Rhythm. Regular and successive harmonious movements.

Rickets. See Rachitis.

Rigor. Sensation of cold, with involuntary shivering; chill.

Rima. Fissure; cleft; furrow.

Rima glottidis. The slit or opening between the vocal chords.

Rolando. An anatomist of Piedmont, Italy (about 1775), who described the so-called *fissure of Rolando*.

Ruga (*pl.* rugæ). Fold or wrinkle.

S

Saccharin. A white crystalline powder derived from coal tar, equal in sweetness to about 300 times its weight of cane-sugar; relates, also, to saccharum.

Saccharum. Sugar.

Saccule. A small sac.

Sacral. Pertaining to the sacrum.

Sacrum. Wedge-shaped bone formed by the union of five vertebræ, attached to the last lumbar vertebræ and coccyx, and forming part of the pelvis.

Sagittal. Resembling an arrow.

Sanguis. The blood.

Sanious. Thin, bloody matter of an ulcer.

Saphena (saphenous). Superficial, such as the saphenous vein.

Saponify. To convert into soap.

Saprophytes. Minute organisms which can only grow in dead or decaying matter; schizomycetes.

Sarcoid. Resembling flesh.

Sarcolemma. The delicate, colorless sheath surrounding the muscle-fibre.

Sarcoma. Fleshy tumor, with either spindle-shaped, round, or giant cells.

Scala. Stairway; name given to the spiral passage-ways within the cochlea.

Scaphoid. Boat-shaped.

Scarf-skin. Epidermis.

Schizomycetes. Parasitic schizophyta devoid of chlorophyll, including all minute organisms known as *bacteria, microphytes, microbes,* and other forms.

Schizophyta. Common name of both groups of schizomycetes and algæ.

Sciatic. Relating to the ischium.

Scirrhus. A firm, hard, carcinomatous tumor.

Scrofula. Predisposition to hard, indolent, glandular tumors, often suppurating slowly and imperfectly, healing with difficulty, and usually occurring on the neck, behind the ears, and under the chin.

Scurvy, or scorbutus. A morbid condition marked by inflamed gums, loosened teeth, hæmorrhagic mucous surfaces, purpuric eruption, and depression, with general anæmia.

Sebaceous. Pertaining to sebum.

Sebum. Suet; a soft, white, fatty, oily substance secreted by sebaceous glands.

Section. A cut; severance; division.

Segmentation. Separation; formation or division into segments.

Semilunar. Crescent-shaped; halfmoon-shaped.

Senile. Pertaining to old age; feeble.

Sepsis. Infection of putrefactive poison.

Septicæmia. Disease aggravated by absorption of pus or putrid matter.

Septum (*pl.* septa). Partition; separating wall.

Serous. Pertaining to serum, or serous membrane.

Serous membranes. Membranes lining closed cavities secreting serous fluids.

Serrated. Toothed, like a saw.

Serum. Watery portion of animal fluids.

Sigmoid. Resembling the Greek letter "S," called sigma.

Sinew. Tendon.

Sinus. Cavity with narrow opening.

Somnambulism. Walking or other actions while asleep, with consciousness suspended, but with the mind and other faculties active.

Sopor. Deep, profound sleep.

Soporific. Inducing sleep; hypnotic.

Spasm. Sudden, irregular and involuntary contractions of muscles, due to reflex action of the spasm-centre in the medulla oblongata above the ala cinerea.

Sphacelus. Gangrene of soft tissue, with complete death of the part. (See Mortification.)

Sphenoid. Wedge-shaped.

Sphincter. A muscle that surrounds and closes an orifice by contraction of its edges.

Sphygmograph. An instrument for measuring or recording the rate, force, extent, and variations of the pulse.

Sphygmus. Pulse.

Spinous. Shape of a spine or thorn.

Spirillum. Spiral schizomycetes.

Splanchnic. Relating to the viscera.

Splenic. Pertaining to the spleen.

Spontaneous. Evolved by itself, without any assistance or any manifest cause.

Squama (*pl.* squamæ). Scale or thin scab.

Squamous. Scaly.

Stasis. Stagnation without morbid condition.

Stenosis. Narrowing or constriction of an orifice or canal.

Sterile. Not productive; barren.

Sthenic. Strong; opposed to asthenic.

Stimulus. That which excites or arouses energetic action.

Stratum. Layer.

Stria (*pl.* striæ). A channel, groove, or furrow.

Stroma. Tissues that constitute the groundwork of an organ.

Struma. Scrofula.

Strumous. Scrofulous.

Stupor. Diminished activity of the intellectual faculties; insensibility; lethargy.

Sty, styan, or stye. Hordeolum; abscess of the eyelid.

Styloid. Pointed; shaped like a pin or peg.

Subarachnoid. Beneath the arachnoid.

Subclavian. Under the clavicle.

Subcutaneous. Under the skin.

Submucous. Part of the membrane just beneath the mucous portion.

Sudor. Sweat; perspiration.

Sudoriferous. See Sudoriparous.

Sudoriparous. Producing or secreting perspiration.

Suet. Sebum.

Suffusion. Spreading or flowing over; slightly diffused congestion.

Sulcus (*pl.* sulci). A groove or furrow.

Superciliary. Situated or being above the eyebrow.

Suppuration. Formation of pus.

Supra. Above.

Supraclavicular. Above the clavicle.

Suprarenal. Above the kidney.

Sylvian fissure. Deep, narrow groove or furrow (sulcus), parting the anterior and middle cerebral lobes.

Synchondrosis. Union or articulation of bones by intervening cartilage.

Syncope. Fainting or swooning.

Synovia. Fluid secreted by a synovial membrane.

T

Tabes. Wasting away; emaciation.

Tabes dorsalis. Wasting of the posterior columns of the spinal cord, producing locomotor ataxia.

Tænia. A tape; tapeworm.

Taliacotian operation. Rhinoplasty; an operation performed by Dr. Taliacotius, an Italian surgeon, in 1575.

Taurin. A colorless, crystallizable substance in bile.

Temperament. The condition of the system with reference to either one of the four forms of disposition, or their combination, viz.: bilious, lymphatic, nervous, or sanguinous.

Temporal. Relating or belonging to the temple.

Tendinous. Pertaining to tendon.

Tendon. Cord or sheet of condensed fibrous tissue connecting either muscle to muscle or to bone.

Tenesmus. Frequent and painful straining and sense of desire for defecation or micturition, generally without discharge.

Tentorium. A process of the dura mater across the back part of the cranial cavity separating the cerebellum from the cerebrum.

Tessellated. Formed into squares; checkered.

Tetanus. Spasms, with rigidity in paroxysms of tonic convulsions of the muscles of the jaw, spinal region, and limbs.

Texture. The interweaving or arrangement of the tissues of an organ; tissue.

Therma. Heat.

Thoracic. Relating to the chest.

Thorax. The breast chamber, formed by the ribs and costal cartilages, spinal column and sternum, between the neck and abdomen.

Thrombus. A blood-clot formed during life in a vessel or tissue.

Thymus, or thymus gland. An oblong, bilobate, soft, glandular body, very variable in size and color, large in the foetus and child, but small or not apparent in the adult, situated in the upper separation of the anterior mediastinum, containing a milky fluid. Its function is unknown.

Thyra. A gate.

Thyro. Prefix referring to the thyroid cartilage.

Thyroid. Shield- or wedge-shaped.

Topical. Local.

Tormina. Twisting, griping pains in the bowels.

Torpor. Numbness; debility; sluggishness.

Toxical. Poisonous.

Trachea. Fibro-cartilaginous air-tube extending from the larynx to the two bronchi; windpipe.

Tractus. Tract.

Transudation. The passage of fluid through tissues, which may collect (sweat-like) in small drops on the opposite surface, or evaporate from it; passage of blood through the vascular walls; oozing through a part.

Trauma. A wound.

Traumatic. Relating to a wound or injury.

Trichina (*pl.* trichinæ). A minute, hair-like worm.

Trichina spiralis. The trichina found in muscles of animals, especially in swine.

Tricuspid. Having three points.

Trifacial. Term applied to the fifth pair of cranial nerves; trigeminus.

Trigeminus. Threefold; the fifth pair of cranial nerves.

Trismus. Lockjaw; tetanus limited to the neck and jaw muscles.

Trochlea. A pulley; a surface grooved like a pulley.

Trochlear nerve. Fourth cranial nerve (patheticus); the motor nerve for the superior oblique muscle of the eye.

Trophic. Relating to nutrition.

Tubercle. Small, rounded eminence.

Tumefaction. Act or process of swelling.

Tumescence. Process of blowing up or swelling.

Tumor. A circumscribed abnormal new formation of tissue.

Tunica. A coat, or membranous envelope.

Turgescence. Swelling.

Tympanum. Cavity of the middle ear.

Tyrotoxicon. Micro-organism; cheese poison.

U

Ulcer. A solution of continuity of soft parts, with loss of substance, production of granulation-tissue and secretion of pus.

Umbilicus. The navel.

Unciform. Hook-shaped.

Uncinate. Hooked.

Unicellular. Composed of only one cell.

Unilateral. Relating to one side only.

Unipolar. A centre-cell with but one prolongation.

Uræmia. Accumulation of urea in the blood.

Urate. Compound of uric acid with certain bases, such as salts of soda, potassa, ammonia, and lime.

Ureter. A tube from 15 to 18 inches long, with the diameter of a goose-quill, conveying urine from the kidney to the bladder.

Urethra. A membranous canal conveying urine from the bladder to the external orifice of discharge.

Urinary. Pertaining to urine or the urinal organs.

Uriniferous. Conveying urine.

Uterine. Pertaining to the uterus.

Uterus. Womb.

Utricle. Small sac or cavity in the vestibule of the internal ear.

Uvula. Conical organ at the soft, free edge of, and pendent from, the palate in the median line above the larynx.

V

Vagina. A five-inch tubular canal extending from the vulva to the uterus.

Vagus. Wandering; the pneumogastric nerve.

Valvula (*pl.* valvulæ). A little valve.

Varicose. Pertaining to varix.

Varix. Permanent dilatation of a vein.

Vas (*pl.* vasa). Vessel.

Vascular. Pertaining to or containing vessels.

Vaso-dilator centre. A centre in the medulla oblongata, the stimulation of which produces dilatation of the blood-vessels.

Vaso-inhibitory centre. See Vaso-dilator centre and Vaso-motor centre.

Vaso-motor. Causing contractions of blood-vessels.

Vaso-motor centre. A centre in the medulla oblongata giving origin to the vaso-motor nerve-fibres.

Vein, vena, phlebs. Vessels conveying blood inward toward the heart.

Vena (*pl.* venæ). Vein.

Venesection. Opening a vein; blood-letting.

Venous. Relating to veins.

Ventriculus. Ventricle; a cavity.

Vermicular. Resembling a worm or its movement.

Vermiform. Worm-like.

Vesica. A bladder.

Vesical. Pertaining to a bladder or cyst.

Vesicle (Lat. vesicula). A small bladder, generally containing fluid.

Vesico-vaginal. Relating to both bladder and vagina.

Vestibule. Entrance. In German called *Vorhof*.

Villous. Provided with villi.

Villus (*pl.* villi). Velvet-like; fine, hair-like projections, especially of the intestinal mucous membrane.

Virus. A substance, the result of a morbid process, capable of producing disease when inoculated; the active agent in the production of any infectious disease; a poison.

Viscid. Sticky; adhesive.

Viscus (*pl.* viscera). The organs contained in any of the three great cavities— cranial, thoracic, and abdominal, especially the two latter.

Vis vitæ. Force of life; the vital power and its effects.

Vital. Relating to life.

Vitellin. The principal proteid in the yelk of egg.

Vitelline. Relating to the yelk of egg.

Vitelline membrane. A very thin cell membrane of the ovum.

Vitellus. Yelk (or yolk).

Vitreous. Glassy; hyaline.

Volar. Relating to the palm of the hand.

Volition. The act of willing or choosing.

W

Woorara. A very destructive poison of Guiana, which contains strychnia; curare; a South American arrow poison.

X

Xiphoid. Sword-shaped.

Y

Yellow elastic tissue. Certain connective tissue fibrillæ, in thickness up to about 11 mmm, anastomosing with each other, found especially in the inner coat of arteries and lining of air-passages.

Z

Zona. A girdle; envelope.

Zone of Zinn. The anterior thickened portion of the hyaloid membrane at the margin of the crystalline lens.

Zyme. Ferment.

Zymogenic. Producing fermentation.

Zymosis. Fermentation.

INDEX.

JUST OUT.

OVER

3000 QUESTIONS

ON

Laws of the Human Body

BY

Prof. J. P. Schmitz, M. D.,

Author of "Human Physiology, Analysis and Digest;" and "The Cause of Diphtheria and the Difference between Diphtheria and Croup."

This work has been designed that the student might better understand the real point contained in the text-book, **"Human Physiology, Analysis and Digest."**

A student often reads a sentence in the book on physiology, and remains in doubt as to the real point contained therein, or perhaps he does not see anything of importance in the sentence.

It will be found that the questions in this work contain really the gist of the various texts, and that the different sentences of the texts contain the **Answers to the Questions.** It will also be noticed that the answers (sentences in the text) on physiology **Follow One Another** in the same order as the questions. The student, therefore, can with ease understand and truly comprehend the laws governing all the organs of the human body. This work, in connection with the author's physiology, is, if I may so express it, a chewing up of the scientific food for more easy digestion by the student.

This book (paper cover) will be sent to any address postpaid for ONE DOLLAR.

Address **J. P. Schmitz, M. D.,**
3321 Twenty-first St., San Francisco, Cal.

www.ingramcontent.com/pod-product-compliance
Lightning Source LLC
Chambersburg PA
CBHW021404210326
41599CB00011B/1009